HARVEST THE WIND

HARVEST THE WIND

America's Journey to Jobs, Energy Independence, and Climate Stability

PHILIP WARBURG

BEACON PRESS, BOSTON

To Tali and Maya

May your generation continue the journey

Beacon Press
25 Beacon Street
Boston, Massachusetts 02108-2892
www.beacon.org

Beacon Press books
are published under the auspices of
the Unitarian Universalist Association of Congregations.

Printed in the United States of America

15 14 13 12 8 7 6 5 4 3 2 1

This book is printed on acid-free paper that meets the uncoated paper
ANSI/NISO specifications for permanence as revised in 1992.

Text design by Wilsted & Taylor Publishing Services

Library of Congress Cataloging-in-Publication Data
Warburg, Philip.
Harvest the wind : America's journey to jobs, energy
independence, and climate stability / Philip Warburg.
p. cm.
Includes bibliographical references and index.
ISBN 978-0-8070-0107-3 (alk. paper)
1. Wind power plants—United States. 2. Wind power
industry—United States. I. Title.
TK1541.W36 2012
333.9'20973—dc23 2011039420

CONTENTS

vii Introduction

1 CHAPTER ONE *Cloud County Revival*

19 CHAPTER TWO *Early Adopters*

39 CHAPTER THREE *Rust Belt Renewables*

59 CHAPTER FOUR *The Chinese Are Coming*

74 CHAPTER FIVE *Working the Wind*

96 CHAPTER SIX *The Path to Cleaner Energy*

116 CHAPTER SEVEN *Birds and Bats*

134 CHAPTER EIGHT *The Neighbors*

161 CHAPTER NINE *Greening the Grid*

180 EPILOGUE

185 Tables

190 Acknowledgments

197 Notes

218 Selected Bibliography

232 Index

Introduction

THIS BOOK IS BORN OF HOPE and frustration. I was an idealistic teenager when millions of Americans gathered for the first Earth Day in April 1970. At public rallies, on college campuses, and in high schools like mine, people demanded an honest reckoning with how we were destroying our environment. It was a time of tumult, a time of questioning.

Then came the Arab oil embargo in 1973, which created panic at the gas pumps and set Washington abuzz with debate about the need to wean America off foreign oil. After college, I joined the energy staff of Senator Charles "Chuck" Percy of Illinois, a former corporate CEO and a Republican who took a pragmatic, bipartisan approach to promoting energy conservation and fossil fuel alternatives. President Jimmy Carter was a willing ally, laying out an energy program aimed at providing 20 percent of America's energy needs from solar, wind, and other renewable resources by the year 2000. As far-fetched as that goal seemed at the time, Congress got behind a number of measures to support investments in renewable energy projects, including wind farms.

My own, as well as the nation's, hopes for a clean energy future were dashed in 1981, when Ronald Reagan took office and immediately set about dismantling the Carter blueprint. Federal funding for renewable energy was slashed as the new president gave preferential treatment to coal, gas, oil, and nuclear development. Wind energy had begun to take hold in California, but new projects came to a near-halt by the mid-1980s—about the time that Reagan symbolized his disdain for renewable energy by having Carter-era solar panels ripped off the White House roof.

The global momentum behind wind power shifted to Europe during the 1990s, with Denmark, Germany, and Spain leading the way. That decade was also a period when climate change moved out of scholarly publications and into the public arena. It was fast becoming clear that curbing our dependence on fossil fuels was an environmental and humanitarian imperative, not just a matter of national security.

After working on environmental and human rights issues in the Middle East for many years, I returned to my native New England in 2003 to serve as president of the Conservation Law Foundation, an environmental advocacy group concerned with energy policy and climate change. From that new vantage point, I was struck by how little progress America had made toward Jimmy Carter's long-forgotten renewable energy target. Wind energy that year accounted for a tenth of a percent of America's power generation, and solar energy even less. Though George W. Bush had supported wind energy as the governor of his home state of Texas, as president he sent an unambiguous signal to the nation and the world: America was unwilling to lead, or even be part of, a serious global effort to curb greenhouse gas emissions.

Stymied by inaction in Washington, my colleagues and I pushed for saner energy policies and practices in New England. While we helped launch the nation's first multistate effort to curb carbon emissions via a "cap-and-trade" regime called the Regional Greenhouse Gas Initiative, we had a much tougher time moving renewable energy forward on a scale that would make a difference. In advocating for America's first offshore wind farm, Cape Wind, to be built near the Massachusetts coast, we found ourselves at loggerheads with an extremely well-funded opposition group that was determined to block any encroachment on their backers' ocean views. What made matters worse, Senator Ted Kennedy—long a champion of renewable energy—had come out against Cape Wind because it was to be sited about five miles from his family's Hyannis Port vacation compound. If the likes of Senator Kennedy were so unwilling to look beyond their own backyard interests, I wondered how and where America could make wind energy happen.

In 2009, I set out to explore this question. My journey began on the Kansas prairie, in half-forgotten Cloud County, where old-timers have long watched their children and grandchildren leave family farms and ranches in search of more lucrative and perhaps less arduous work elsewhere. A large commercial wind farm went into operation there in December 2008, its giant turbines generating enough electricity for 55,000 Kansas and Missouri households. To understand what wind power has brought to this struggling corner of rural America, I spent time getting to know the landowners who have rented out sections of their fields and pastures, the educators who have reshaped the local community college as a training center for wind energy technicians, and the developers who worked with community partners to move the Meridian Way Wind Farm from concept to completion.

I also traveled to Rust Belt towns of the Midwest, where I learned about the thousands of new jobs that the wind industry has created for young and old alike, in retooled factories that once made car parts, printing presses, and construction cranes. And at wind farm sites across the heartland, truckers and construction crews expressed their excitement at being part of an emerging industry with a promising future, and newly trained technicians described the rigors of working atop 300-foot towers in hundred-degree heat and subzero cold.

Wind is not just a new presence on the American working landscape. It is also an expanding force in the global marketplace. To gain a better grasp of wind's broader market dynamics, I traveled to Denmark, which is proud to be producing one fifth of its power from wind. This small Scandinavian nation is also home to the world's top-ranked turbine supplier, Vestas Wind Systems, which made the machines that now spin out Cloud County's wind power. My next stop was China, vast and ambitious, outstripping all other nations in its wind energy use. Not surprisingly, the Asian giant is also emerging as a major rival to Europe and the United States in its wind energy manufacturing.

Today, America gets about 3 percent of its electricity from wind—hardly enough to get excited about. Yet, in raw terms, our wind energy resource—like our solar potential—vastly outstrips our current and future power needs. The U.S. Department of Energy conservatively predicts that, by 2030, wind energy could provide 20 percent of America's electricity. Other experts see wind generating more than half our nation's electricity by mid-century, so long as we invest in new transmission lines that will open up some of the remote areas where our wind resources are greatest, and so long as we develop a "smart grid" that can match the variable flow of wind-generated power to consumers' needs.

Like any other energy technology, wind power has its downsides. Birds and bats may be killed by turbines, and the habitats of closely watched species like the sage grouse, the prairie chicken, and the Indiana bat may be at risk in some areas. Wind farms also can pose problems for people living nearby. Some object to the visual presence of towering turbines; others complain about the noise. At places as far-flung as Vermont's Taconic Mountains and the Flint Hills of Kansas, the tensions surrounding wind energy have caused bitter and lasting divisions among neighbors and former friends. Wind power's future success depends on our addressing these issues responsibly.

As I made my way across vast stretches of a country that I previously only knew through books and the media, I often thought of the advice that Franklin D. Roosevelt gave to a headstrong Wall Street banker who had just resigned from his Brain Trust, disillusioned with the New Deal's spending and monetary policies. "Please get yourself an obviously secondhand Ford car," the president counseled. "[P]ut on your oldest clothes and start west for the Pacific Coast, undertaking beforehand not to speak on the entire trip with any banker or business executive (except gas station owners), and to put up at no hotel where you have to pay more than $1.50 a night."[1] The banker—my father—apparently felt no need to follow FDR's advice. I think he missed out on a lot.

Although the motels where I stayed charged considerably more than $1.50 a night, the cross section of Americans whom I met—

farmers, ranchers, factory workers, shop owners, truckers, crane operators, and more—deepened my appreciation and respect for the people who inhabit so much of this land. I also came to understand much more fully the ways that America's push toward renewable energy has changed people's lives for the better. Tens of thousands of Americans have found new jobs in the factory and field. Thousands of others have seen their flagging farm incomes boosted by generous, multi-year lease payments from wind developers.

Yet these people see more in our nation's wind energy investment than a pay stub or a rent check. The devastating consequences of climate change may still seem remote, but the people involved in our wind energy sector are fiercely proud of their role in advancing the nation's quest for energy independence. Too many of their family members, friends, and neighbors have been sent overseas to fight wars that they see, at least in part, as a struggle for control over waning energy resources. Too many loved ones have come back from those wars defeated in spirit, physically maimed, or in body bags. Too much taxpayer money has been spent fighting distant wars with no clear purpose—funds that would have been far better spent creating jobs and new industries here at home.

Investing in wind energy will not bring an end to Middle East strife, but it *can* begin to wean our nation off the fossil and nuclear fuels that we have come to associate far too closely with American prosperity. Now is the time to reshape our energy economy. With wind, we can tap an inexhaustible domestic energy resource while showing that America finally is willing to join—and even lead—the battle for climate stability.

Turbine basics: The Vestas V90 3-megawatt turbines shown here are part of the Meridian Way Wind Farm in Cloud County, Kansas. At a well-sited wind farm, 1 megawatt of installed capacity produces enough electricity for roughly 270 average U.S. households. Graphic by Jason Fairchild of the Truesdale Group/Recycled Paper Printing Inc.; photo by Philip Warburg.

CHAPTER ONE

Cloud County Revival

WHEN THE U.S. POSTAL SERVICE commemorated a century and a half of Kansas statehood, those familiar with the state found it easy to fathom the stamp's chosen graphic: an array of windmills against a backdrop of broad blue sky. Even the state's name is drawn from the wind, or more precisely from an Indian tribe that honored the winds sweeping up into the region from the Gulf of Mexico. They were called *Kansa,* or "people of the South Wind."

The wind is at times a frighteningly destructive force in Kansas life. Killer tornadoes have leveled entire towns and have made tornado chasing a precarious summer sport for daredevils seeking grim moments of media attention. Yet it also offers a vast and largely untapped source of clean, renewable energy. According to a recent government survey, Kansas could install sufficient wind power to supply almost 90 percent of America's total present-day power consumption.[1] Wind energy investments in the Sunflower State today have barely scratched the surface of that vast potential. About 7 percent of the state's electricity came from wind in 2010, with coal providing most of the remainder.[2]

The story of putting the wind to good use could be told with slight variations throughout the Midwest and Great Plains states. As far back as the 1850s, Kansans recognized its value as an economic resource. Water-pumping windmills sprouted alongside farmhouses and in cattle pastures, greatly easing the burdens of frontier homesteaders. Initially made of wood, these early machines were easily repaired by self-sufficient farmers and ranchers. By the turn of the twentieth century, steel began to edge out older wooden designs and the maintenance burden gradually shifted to full-time itinerant windmillers. Some wind machines bore whimsical names like Albion, Climax, Eclipse, and Eureka; others had mechanical monikers like the Axtell Ever-Oiled and the I. X. L. Steel. Manufacturers were scattered across the Midwest and Great Plains states, and dozens based their operations in Kansas.[3]

Water-pumping windmills were widely used in Kansas until the late 1940s, when the electric grid finally reached rural sections of the state. Even then, many farmers continued to rely on wind to draw water up out of the ground; their windmills were already in place and, with minor tinkering, offered a free source of power. The remains of these machines still dot the Kansas prairie today. I encountered many of them in my travels across the state: spindly steel-frame obelisks, some standing bladeless, others with a few ragged slices of metal groaning and rattling in the breeze.

A second wave of windmills, or "wind chargers," brought a thin stream of electric power to rural Kansas in the 1920s. Hooked up to a car battery, these crude machines with their two-blade wooden rotors generated just enough juice to let farmers run a radio—a much-valued novelty at the time—or a few light bulbs. Older Kansans recall how these devices brought the wider world into their lives by opening up the airwaves. Larger wind chargers were also marketed at the time, one of them luring housewives with the promise that with home-grown electricity running their appliances, ironing would become a pleasure and cleaning a joy, but their price—in the hundreds of dollars—was beyond the reach of most farm families. In any case, wired-in electricity to most parts of the state made household-scale wind power obsolete by the late 1940s.

Over the next half-century, cheap electricity from large conventional power plants kept renewable energy at bay in Kansas, as through much of the nation. Kansas largely missed out on the turbulent years of commercial wind energy experimentation that swept through California in the 1970s and early 1980s. Only after the turn of the millennium did wind energy find new life in Kansas. Spurred on by a property tax exemption adopted in 1999 and further bolstered by federal tax credits for renewable energy-based power,[4] smart developers grasped the wind's enormous untapped potential and began to exploit its commercial possibilities.

The state's first large commercial wind facility, the Gray County Wind Energy Center, went on-line in 2001. Its "installed capacity"—the maximum amount of power it can generate at any moment—is 112 megawatts, or 112 million watts. No power plant operates at full capacity on a continuous basis, however. This is especially true for wind farms, which depend on a power source that can vary from hour to hour, day to day, and season to season. The Gray County Wind Energy Center operates, on average, at about 41 percent of its installed capacity—higher than most other wind farms.[5] At that rate, it churns out enough electricity to meet the needs of more than 32,000 Kansas households.

The Gray County project may have been the first big wind farm in the Sunflower State, but others soon followed. One of them was the Meridian Way Wind Farm, sited on a gently rolling stretch of prairie in Cloud County, just about a hundred miles north of Wichita on Interstate 81. The sixty-seven giant turbines at Meridian Way were dramatic newcomers to a predominantly horizontal landscape. Today they are a transformative force in a community that has been in steady decline ever since mechanization overran the local farm economy decades ago.

Kurt Kocher is the fourth generation in his family to raise cattle and till the soil on the family homestead, about nine miles south of Cloud County's sleepy administrative seat, Concordia. He fears he may be the last. "There are simpler ways of making a living than farming," he says, pointing to the trailer where his son Kenton—a high school junior—is sweeping out crop waste. "He thinks it's forced servitude because it's spring break this week!"

Whether Kenton will carry on the family farming tradition is an open question. "He doesn't know what he wants to do," his father grumbles as he adjusts the faded red baseball cap that never leaves his close-cropped, graying head. He knows his son will have plenty of company if he ends up leaving the farm. Over the past half-century, young people have streamed away from Cloud County and other rural communities like it. Some search for steadier work, others want less physically demanding jobs, while still others are fleeing the isolation and tedium of life on the farm. It's not that farming is on the wane. Overall acreage under the plough across Kansas has actually held steady for decades. Yet today, fewer hands are needed to prepare the fields and harvest the wheat, soybeans, sorghum, and sunflowers that have long been the area's staples.

Farms may have gotten bigger, with fewer people tending them, but not much else has changed in Cloud County for a very long time. Not much, that is, until the U.S. subsidiary of a Portuguese electric utility targeted the county for wind development. Nine of Meridian Way's wind turbines now soar above the Kochers' cattle pastures and grain fields, bringing in rent for the placement of each turbine, additional rent for the land where a transformer station now stands, payments for access roads and power-line easements, and a royalty on every kilowatt generated by the project. Kurt didn't talk numbers, but I could surmise from what I knew about landowner compensation elsewhere that the Kochers' wind-generated income must be closing in on six digits.

The income stream from wind may or may not help keep Kenton Kocher and his younger brother, Tyler, on the family farm, but it certainly brightens Kurt's outlook on the future. Crops can fail, cattle prices can plummet, but the wind blows steadily across the Kansas prairie, and Meridian Way's developers have promised to tap it for at least a quarter-century to come.

Kurt is, above all else, a practical man. You can see it in the soil stains that are ground deeply into his jeans. And you can hear it in the unsentimental way he talks about his prized herd of 150 Red and Black Angus cattle. "These won't be hamburger," he tells me as he chases four young calves out of a trailer, setting them out to pasture for the summer. By early fall, they will be sent to a commercial feedlot, and by early winter, the more "efficient" among them will be turned into prime cuts of beef little more than a year after their birth. "They're not an endangered species," he assures me.

Kurt regards Meridian Way's wind turbines in the same unadorned, pragmatic way he thinks about his Angus cattle and his fleet of farm machinery. A slight smile emerges from beneath his graying goatee as we look out over a large pasture where a half-dozen turbines face due south into the wind. "For our business, we like to know which way the wind is blowing if we're applying weed killer or something along those lines," he says. He thinks of the wind turbines as very large weathervanes. Each turbine takes up no more than half an acre of ground, so he is free to continue tending his fields and grazing his cattle below them.

Helen Kocher, like her son, is little bothered by the turbines. Her eyesight may be dimmed by cataracts, but her hearing remains sharp enough to register the swishing sound of the turning blades, which she says is especially noticeable in the evening, when all else is quiet. It reminds her of waves brushing up against a ship—an unfamiliar sound on the Kansas prairie, but one she has encountered on her travels over the years. "I don't know if it's the speed of the wind or the direction, but you get some pretty big waves sometimes!" Usually, though, she finds the sound just blends in with the sounds of working life on a bustling prairie farm. The steady flow of funds from the wind farm's developers makes it easier for the Kochers to put up with a few minor annoyances here and there.[6]

Ray Mason is another landowner who has made an easy adjustment to wind energy's arrival in Cloud County. Now in his early sixties, Ray works part time managing the American Legion men's club in Concordia. It was about ten o'clock on a Thursday morning when I arrived at the Legion's club hall, a building clad with sky-blue aluminum siding just off West Sixth Street, the spine of Concordia's four-block downtown business district. Toward the back of a dimly lit barroom, acrid from years of accumulated tobacco smoke, a few older men in checkered shirts and blue jeans sat playing cards at a small table. They looked up briefly as I entered. Ray, who stood beside them, gave me a warm smile before offering me his firm farmer's handshake. He then led me around the bar to a game room, slightly better lit, where we pulled two chairs up to an empty poker table.

Born in Concordia, Ray grew up on a farm just east of town and worked as a farmer and rancher until a few years ago, when he decided he needed a rest. He first encountered Meridian Way's developers back in 2003, and it quickly became clear that about five hundred acres of the family's upland pasture were good prospects for wind turbines. When he was offered several dollars an acre for a five-year option to develop wind on the land, the whole project seemed so farfetched that Ray doubted it would ever happen. He spoke with a few neighbors, though, and together, they agreed to get on board. Then they waited.

A few years into the option period, Ray began to get itchy. Next time the "wind boys" came to town, he laid the bait. "I've been waiting on you guys for so long," he told them. "Hell, I'm just gonna sell that damn ground down there. I'm not gonna mess around any longer."

His ploy worked. "One younger guy got out his map, and I showed him where I was. He folded the map and looked at me, and he said, 'Don't you sell that land to anybody unless you sell it to me!' So I knew right then that we was gonna get at least one turbine."

In fact, Ray ended up with five. "I was probably as shocked as anybody," he told me. Of the five turbines slated for his property, two were on 120 acres of marginal grazing land that his father bought for only $25 an acre back in 1957. "It was a rough old farm, you know, with rocks kinda close to the top of the ground and short grass, overgrazed." Ray shook his head, and his eyes welled with tears as he thought of how pleased his father would be if he knew that rugged patch of land would one day bring in $20,000 a year, most of it from wind. "He wouldn't believe it," Ray sighed.

Combining his per-turbine earnings with other payments for access roads, Ray's annual cash flow from the wind farm must be close to $50,000, a welcome subsidy that will eventually benefit his children and—if the wind farm stays in service long enough—his grandchildren. Ray's family is a strong advertisement for community continuity, defying the trends driving younger people out of Cloud County and into areas with richer job prospects. All of his children live in Concordia, where his son works in a local metal shop and one of his daughters is a registered nurse.

While he can't force his children or grandchildren to work the land, Ray is doing what he can to keep them tied to the family homestead, as landlords if not as farmers. "I seen my folks work for that sucker pretty hard. I worked pretty hard too, so I don't want someone to just piss it away." He recently created a trust to maintain his family's ties to the land, at least for another generation or two. "I got it set up so they can never mortgage the land or sell it," he says. "So it'll pass on to my kids, and then to my grandkids. Then I quit."

Meanwhile, Ray is happy to have the extra income that the wind farm is sending his way. "Two months from now, I'll be sixty-two, and it fits in good to my retirement, y'know."[7]

The man responsible for lining up the land deals for Meridian Way was Jim Roberts. An Oklahoma native, Jim was first dispatched to Cloud County in the spring of 2003 by Selim Zilkha and his son Michael, co-owners of Zilkha Renewable Energy in Houston. The Zilkhas had already proven themselves as highly successful energy entrepreneurs. When they sold their oil and gas holdings in the Gulf of Mexico for a billion dollars in 1998, the acreage they held reportedly surpassed all other prospectors operating on the Gulf's continental shelf. Their smart use of 3-D seismic data had yielded a drilling strike rate that was double the industry average.[8] And now the Zilkhas were looking to get into wind.

Before joining up with the Zilkhas in 2001, Jim had worked for many years in the oil and gas industry. There he honed the diverse and subtle deal-making skills that are essential to negotiating leases for energy projects. His wife and family live in Edmond, Oklahoma, but

Jim spends much of his time on the road. As soon as the land needs for one project are nailed down, he's on to the next, living out of motel rooms and trailers and eating most of his meals far from the family dinner table. He loves his job, though, and is happy to be promoting an alternative to the fossil fuels that he helped suck out of the ground for so many years. "I believe in the need for clean, renewable energy," he told me as he began to recount his role in moving the Meridian Way Wind Farm from dream to reality. "I'm very proud about my work in this industry."

The first wind project that Jim lined up for Zilkha Renewable Energy was the Blue Canyon project, now the largest in his home state. The company then sent him to scope out prospects for a wind farm in the Flint Hills of Kansas. To Jim and other wind developers, the Flint Hills were alluring. Located in the eastern part of the state, they are relatively close to Kansas City and other population centers. Just as important, winds in the Flint Hills are the strongest in the state. But Jim and his Zilkha colleagues were in for a beating. An alliance of ranchers and conservationists soon made it very clear that wind turbines were a bad match for one of the state's rare remaining expanses of untrammeled tallgrass prairie. Audubon of Kansas stood firmly against any wind farms in the area, and it was joined by state leaders of another leading conservation organization, the Nature Conservancy.

Jim's normally soft Oklahoma drawl took on a sharper edge when speaking about Flint Hills anti-wind organizers, whom he pegged as NIMBY, or "not-in-my-backyard" activists. One of those opponents was a professor of English at nearby Emporia State University, Jim Hoy. Professor Hoy is an expert in Great Plains folklore, with a special passion for the history of Flint Hills cowboys. When one wind supporter likened the proposed wind turbines to ballerinas, Hoy derided the idea of "scores of 350-foot-tall ballerinas desecrating the native prairie grass that has covered the Flint Hills since the last Ice Age."[9] In January 2004, as public pressures mounted to keep wind farms out of the area, Governor Kathleen Sebelius created a Wind and Prairie Task Force to try to resolve the controversy. While other wind developers put their project plans in neutral and waited for the task force's recommendations, the Zilkha team decided to look elsewhere for a community more welcoming to the company's ambitions. It didn't take long to find friendlier terrain about a hundred miles northwest of the Flint Hills.

Cloud County seemed like a great fit from the start. First, there were the southerly winds that the Kansa tribe had immortalized. Jim Roberts knew those winds would get an extra boost from the updraft off a long, south-facing ridgeline. He also knew that the area's higher elevation, about 1,600 feet above sea level, would give the winds yet a further push. Even before he set up equipment to do more precise measurements, it was clear that the area's climate and topography were in near-perfect alignment.

The land was favorable, too. Well over a century of continuous farming had transformed virtually all of Cloud County into a working landscape. Where the hills were too rugged for row crops, multiple generations of cattle had been set to graze. Greater prairie chickens—grassland birds whose habitat has been decimated throughout most of the Midwest—were still sighted occasionally, but Jim felt confident that measures to protect the breeding and roosting grounds for these birds could be developed.

Every bit as encouraging as the land and the winds were the people. The locals struck Jim as "salt of the earth"—plainspoken and straightforward, with a no-nonsense agrarian pragmatism and a hands-off attitude toward government regulation. Unlike many other Kansas counties, Cloud County has no rural zoning, giving landowners a relatively free hand in deciding what to allow on their property. "From the time we got there, we knew we were going to be accepted," he recalls.

Despite his initial enthusiasm, Jim proceeded with the canniness and caution that are the stock of the land man's trade. "Stealth" is the word he uses to characterize his first steps in scoping out a possible wind farm site in the county. Having just retreated from the Flint Hills melee, he wanted to be sure he had core supporters among local landowners before going public with the project. He was also wary of competitors, one of whom had already put up a meteorological tower to collect more precise wind-quality data. "We're not going to go in there with flags flying and whistles and alarms going off because it's a very competitive industry," he acknowledges.

Describing his next steps, Jim reveals no sources and names no names. "We first made contact with a landowner," he tells me. "We had strategically selected this person because of the land mass that this gentleman had, and we did some research and found he was kind of a ringleader." From that initial contact, the circles of consultation grew—quietly and selectively. The landowner whom Jim had targeted for his opening gambit soon brought in just the catalyst Jim needed for the next stage: Kirk Lowell, a farmer-turned-business-booster who heads up CloudCorp, the county's economic development corporation. Kirk had already identified wind as a major economic growth prospect for the county, so he was more than willing to help out. A wind energy steering committee checked out Zilkha's credentials at Kirk's behest. When word came back that the company was legit, Jim knew it was time to reach out to other landowners.

Jim called a meeting for invited landowners only. He laid out a preliminary proposal for developing a wind farm on agricultural land about eight miles south of Concordia. In the first stage, he explained, participating property holders would be paid to give Zilkha a multi-year exclusive option to explore wind energy prospects on their land. If all went well, smaller parcels would later be leased for wind turbines, access roads, underground cables, and other infrastructure associated with the farm.

Jim teamed up with another Zilkha land agent. Working with a rapidly growing circle of landowners, the two of them assembled a huge swath of land for possible development: 22,000 acres in all. In six weeks' time, they had seventy landowners on board.

The Zilkha crew then geared up for the next stage of development: working their way through a dizzying array of studies and permit applications. Meteorological towers were erected to get a more accurate read on local wind patterns and climate conditions. The Federal Aviation Administration needed to sign off on the project, ensuring that the turbines would not endanger planes taking off and landing at Concordia's airstrip. The Department of Defense had to be satisfied that its radar installations in the area would continue to function unimpeded. The Federal Communications Commission was involved in protecting local radio airwaves.

On top of those issues, there was a lot of work to be done to protect the local environment. A joint project was launched with state conservation groups to set aside land for the mating and roosting of greater prairie chickens; protocols were prepared to protect the area's dung beetle population; and measures were developed to preserve low-lying wetlands in compliance with state and federal law.

Jim knew it was time to name the baby. A four-lane divided highway may seem an improbable inspiration for the naming of a forward-looking renewable energy project, but Interstate Highway 81 was exactly that. The highway runs right through the project area; its precursor was a dirt road that tracked just west of the Sixth Principal Meridian, the north-south map line used by government surveyors to lay out U.S. territories under the Kansas-Nebraska Act of 1854. That old thoroughfare was a lifeline connecting Cloud County farmers and ranchers to southern markets in Salina, Wichita, and beyond—much the same role that Highway 81 serves today. People called it Meridian Road.

"Meridian Way" was the name Jim and his team gave to the wind farm. In crafting the project's logo, artists were asked to link the excitement and promise of twenty-first-century technology to the aura of this historic road. They came up with a simple green graphic showing a highway with a dotted median converging on two tall wind turbines in the distance.

With the project named and the permitting process well under way, most of the pieces were falling into place. Before the wind farm could be built, though, two major challenges remained: finding a guaranteed buyer for the wind farm's electricity and raising the capital to build it.

The day was fast approaching when Zilkha would have to put real money on the table to buy and install the turbines and other equipment needed to turn nature's gift into a marketable commodity. In the old days, farmers might spend a few dozen to a few hundred dollars to buy a windmill. Today's giant wind turbines are in a different league entirely. Meridian Way ended up costing the developers $340 million—an average of $1.7 million per installed megawatt for the 201-megawatt wind farm.[10]

Facing that anticipated financial burden while carrying the costs of several other wind farms, the Zilkhas decided they needed deeper pockets to carry the company's mission forward. In March 2005, they sold a controlling interest in Zilkha Renewable Energy to the then-robust global investment banking house, Goldman Sachs, which changed the company's name to Horizon Wind Energy.[11] Only two years later, Goldman flipped its investment, selling the company—valued at $2.2 billion—to Energias de Portugal (EDP), the leading electric utility in Portugal and a huge player in the Iberian peninsula's natural gas sector.[12] EDP, in turn, assigned Horizon to its renewable energy division, headquartered in Madrid.

With the substantial backing of Goldman Sachs and then of EDP Renewables, Horizon was able to take the next crucial step in advancing the Meridian Way project: finding a long-term buyer for the wind farm's electricity. Converting wind energy to electricity was one thing—that was Horizon's job. Carrying the electricity to the households and businesses that needed it was a whole different story. For that, Horizon needed buy-in from a utility, ideally one with transmission lines already in the area. Lining up a power purchase agreement is the final threshold that shifts a wind farm's developers from paper-pushing to earth-moving.

The quest for a power purchaser led to a surprise development. In the fall of 2007, just as Empire District Electric, a Missouri-based utility, declared its willingness to buy 105 megawatts of Meridian Way's power, a second prospective energy buyer emerged: Westar, the largest electricity distributor in Kansas. Jim Roberts jumped at the challenge, shifting into super-high speed to assemble a second land package that could double Meridian Way's output.

Once again, Jim needed a ringleader who could help him work with landowners to put together another large, contiguous land parcel. Kurt Kocher was his man. Kurt commanded the respect among his peers that was vital to Jim's success. "He kind of became the trumpeter of our praises," Jim recalls. "He farms and ranches responsibly. He's a hard worker. He's got a good operation. He's well-known. . . . You like to get those people in your project up front." Jim pauses. "I'm kinda giving away a little of our strategy here, but it's a no-brainer for any wind company. You're going to try to get your best landowners first."

Within just a few months, Jim's alliances with Kurt and others landed Horizon just what it needed to bring Westar into the deal. Wind development rights on another 12,000 acres were now in the company's hands.[13]

By December 2007, Horizon had all the important preliminaries in place: the wind, the permits, more than a hundred local landowners on board, and commitments from not one but two power purchasers. The company now geared up for the final phase of project development: building the wind farm itself. Horizon wanted the wind farm up and running by the end of 2008, which was essential if it was to tap a 30 percent federal production tax credit that was due to expire on December 31 of that year.

Enter Carole Engelder, a highly skilled, no-nonsense project manager who exemplifies many of the workers who have recently joined this frontier industry. A former employee of Amoco, BP, and a chemical startup, Carole came with a quarter-century of experience overseeing engineering, logistics, and procurement for major projects in the more traditional oil and chemical sectors. She had no wind industry background before joining Horizon in 2007, and she was unmoved by any high-minded idealism about freeing America from the stranglehold of foreign oil or the ravages of climate change. In brisk conversation, she sums up the attraction that wind held for her: "It's fast-paced, it's cutting-edge, it's new in this country. . . . The fact that it's wind, I have to admit, doesn't have anything to do with it." What drew Carole to wind was the adrenaline rush of moving a new industry forward.

Moving this particular project forward meant getting sixty-seven wind turbines in place and hooked up to the grid by Christmas 2008. Given the sheer size of these behemoths, this was no small feat. Horizon chose to use three-megawatt turbines manufactured by Vestas Wind Systems, a Danish company that is well established as a world leader in wind energy technology. The Vestas V90 has a three-bladed rotor whose diameter is the length of a football field. Each tower, transported in four segments, is nearly 260 feet tall and weighs 344,000 pounds. It is topped by a hefty mechanical chamber, or nacelle, weighing 154,000 pounds and containing the turbine's gearbox, generator, transformer, and control systems. With blades spinning, the V90 reaches 410 feet in the air—just about the height of a forty-story building.[14]

Fitting together the many pieces of the Meridian Way puzzle in just ten months required what Carole describes as "choreography." Like a stage manager scrambling before the curtain goes up, she and on-site project manager Alvin Cargill had to make many big decisions and a myriad of smaller ones. They needed to recruit a full complement of contractors and subcontractors, some three hundred workers in all, each with their specialized trade, all of them demanding careful sitewide coordination. They needed precise locations for every tower, transformer, and power distribution line. They needed a long list of crucial components delivered to each turbine site. Access roads were a challenge in themselves. Some had to be planned and surfaced from scratch. Many of Cloud County's existing roads had to be widened and reinforced with concrete to withstand the weight of the massive, specially engineered semitrailers with their giant cargo of towers, gearbox assemblies, blades, and transformers.

The work required not only precision but flexibility. If a weather report showed an approaching thunderstorm, the dozen people poised to lift a multi-ton turbine tower segment into place would have to set down tools and stand clear of their 300-foot-high crane. With no grounding, this machine is its own lightning rod. "Perfect planning isn't about planning the perfect project," Carole tells me. "It's about perfect contingency planning."

With so much happening in a few short months, disruptions to the daily routines of local farmers and ranchers were inevitable. Skilled at orchestrating the logistics of a complicated and fast-paced construction job, Carole was less patient in dealing with disgruntled neighbors. "Gnats," she calls them.

"N-A-T-S?" I echo, thinking this is a wind industry acronym.

"*G–n–a–t–s*," Carole spells out. "Annoyances. Alvin takes care of them so I don't have to hear about it."[15]

One of the "gnats" Jim Roberts and Alvin Cargill had to contend with was Bonnie Sporer, a landowner who signed up to allow wind towers on her property. Bonnie, a seasoned rancher in her eighties, is a canny businesswoman. She's a Daughter of the American Revolution, proud of it, and equally proud to be the descendant of a Civil War doctor who brought his family to Cloud County in a covered wagon shortly after the Union victory. Farming and crop management have run in Bonnie's family for generations. Her family has farmed rich bottomland in the Republican River valley, just north of Concordia, for as long as anyone can remember. She grew up on that farm and continued to manage it as an adult, together with other family members. She also spent many years working for the U.S. Department of Agriculture, running the government's crop insurance program in Cloud County and points west—the largest crop insurance program in Kansas, she boasts. She served, too, as the first woman on the county's Soil Conservation Board, and she made sure I knew it.

Bonnie began her relationship with the wind farm developers by driving a hard bargain with Jim Roberts and his land team back in 2003. "They asked if I was interested in leasing some land," she says. "I made them take all of it." That amounted to 1,800 acres of ranchland, for which the company paid her several dollars per acre for the option to tap the available wind. Once the wind farm's layout was finalized, she says she received close to $80,000 for an easement to accommodate underground electric cables, plus payments of over $16,000 a year for two wind turbines sited on her property. On top of that, she is paid rent for access roads to the turbines that have been erected on her land.

Jim won't discuss the specifics of Horizon's payments to Meridian Way landowners, though he allows that Bonnie can sometimes be a little loose with the details. "One minute you'll be having a conversation with her, telling her that it looks like she'll have five wind turbines on her property. (He stresses that this is a hypothetical number.) A half-hour later, someone calls and says, 'We didn't know Bonnie was getting ten turbines!'"[16]

I first encountered Bonnie when she strode into Meridian Way's field office on a warm afternoon in May 2009, on my second visit to Cloud County. She wore rubber boots up to the knee and left a boisterous border collie barking outside the door. Bursting with energy, she reeled off a long list of complaints with the way the construction process had unfolded. In one instance, a construction worker left one of her cattle gates open, she said, and her entire herd had wandered out before anyone noticed. Then those out-of-town greenhorns went racing around in their pickup trucks, trying to round up the creatures as they wandered off in all directions. "That never works!" she exclaimed, shaking her arms in the air for effect. "It only panics the animals." One of the cows—Number 41, she specifically recalls—was never found.

Bonnie recalls sharing dreams with other family members that their unyielding ancestral terrain might surprise them in the future. "We always laughed that someday we'd get lucky and hit oil or something." When Meridian Way's turbines started turning in December 2008, just ten short months after the first foundations were poured, Bonnie knew that, for all the aggravations, this was as close to a gusher as she was likely to see.[17]

Landowners, without a doubt, are the primary beneficiaries of Meridian Way, but there are other gains that reach the broader Cloud County community. During those intensive months of building the farm, about half of the 300 onsite workers were locals.[18] If the farm doubles in size as planned, there will be similar opportunities for local workers, including truck drivers, metal welders, electricians, crane operators, and road construction crews. With all those workers hungry by noontime, local eateries and grocery stores can count on a substantial boost. In 2008, retail sales in Concordia—and therefore local sales tax revenues—jumped a full 24 percent.[19]

Meridian Way has also spawned a cutting-edge training program for wind energy technicians at Cloud County Community College. Bruce Graham, the program's creator and main driving force, has suddenly brought this little-known, two-year college into the limelight as one of the nation's leading pathways to a burgeoning, twenty-first-century industry. Raised on a Cloud County dairy farm, Bruce—a former high school science teacher—finds himself scrambling to keep pace with a program that, in just a few years, has grown from only four registered students to more than a hundred job-hungry enrollees. Until the recession hit, he had a hard time keeping students enrolled long enough to finish their two-year training before they got snapped up by wind developers. Even after the economic downturn cut the rate of new wind development by nearly half in 2010, Bruce still was able to place most of his students in entry-level jobs in the industry.[20]

Along with teaching classes and hustling to raise funds for his program, Bruce finds himself in constant demand as a public lecturer. On an evening when I visited the program's makeshift classrooms in rented space at a Concordia shopping mall, he entertained a room full of aging, tattooed visitors from the Kansas-Nebraska Radio Club. A busload of Nebraska farmers had visited the school the previous week, wanting to know how they could bring wind energy technology and jobs to their own communities.

Bruce's wife, Michelle, is Meridian Way's administrative coordinator. Along with managing contracts and dealing with daily operational issues, she—like her husband—hosts a steady stream of curious wind farm visitors: church groups, schoolchildren, and more than the occasional journalist. Up on all the details, she fields questions deftly and conveys a contagious enthusiasm for the project.

Looking at what Meridian Way means to the Cloud County economy, Kirk Lowell is every bit as bullish about wind as the Grahams. I met the Cloud County economic development director at CloudCorp's modest storefront headquarters, just a few doors away from Ray Mason's American Legion Post. As he ushered me into his office, it became obvious immediately that I was in the presence of a true believer. His polo shirt sported the Meridian Way logo. On his desk was a scale model of the Vestas V90, just like the turbines now operating at Meridian Way. And on his wall was a caricature of Grant Wood's *American Gothic* farm couple, minus the pitchfork, with two tall wind turbines towering in the background.

Kirk knew that I was visiting from the East, and he wanted to make sure I understood the roll-up-your-sleeves spirit that pervades Cloud County. "We have a very common-sense kind of people here. We know that we have to do something as a nation about our energy situation." He looked me straight in the eye. "You can't just talk about it. It bothers people in the Midwest when you say, 'We need to do this, this, and this. *I'm* not gonna do it, but we expect *you* to do it.'"

Though he was too polite to mention it, I suspected that Kirk was alluding to the raging controversy back East over the long-stymied offshore Cape Wind project on Nantucket Sound. What happens to all those enlightened East Coast idealists, he must have been wondering, when tough choices have to be made about where to build new wind farms? His message was clear: "Somebody has to step up to the plate and host these things."

Kirk's commitment to wind is rooted less in concern about climate change than in a weariness with wars fought over oil, thousands of miles away from the orderly calm of his home territory. His logic is elementary: "Either we are going to have to put wind turbines *on* our Kansas prairie or we're going to continue to put our fine young men and women *under* it."

Cloud County is populated by more than the usual complement of hardened realists, but the county has its share of optimists as well. Wind energy alone may not be enough to reinvigorate Cloud County and places like it across rural America. That will take the concerted efforts of many people pulling together, reshaping and rebuilding their communities. There is no doubt, though, that Meridian Way has helped give this small slice of the American heartland a new pride of place and a belief in better things to come.

CHAPTER TWO

Early Adopters

IT WAS HARDLY A SURPRISE that the giant wind machines at Meridian Way came from Denmark. Vestas Wind Systems has long been a wind energy technology leader, with a supply chain reaching back to America's first big wave of wind power development in the 1980s. Denmark has also distinguished itself as a global pioneer in generating its own electricity from wind. A fifth of the country's power today comes from wind, and plans are afoot to raise that contribution to as much as 80 percent of the country's power supply—all part of a broader government plan to free Denmark from fossil fuels and reduce its greenhouse gas emissions by 80 to 95 percent by 2050.[1]

I traveled to Denmark to get a close look at wind turbine production. From Copenhagen I drove to the village of Lem, set on a flat stretch of farmland near the western rim of the Jutland peninsula. This is where Vestas got its start and where some of the company's manufacturing still takes place.

Lone Mortensen, director of people and culture in the Vestas blades division, met me at the entrance to a neat campus of white factory buildings. Appearing wholesome and informal in her slacks and Argyll sweater, Lone (pronounced "LO-neh") had recently returned from a two-year stint in Portland, Oregon, where she helped ramp up the company's North American communications. Despite both countries' refusal to sign on to any greenhouse gas reductions, the United States today ranks second only to China in both its annual installation of wind turbines and its cumulative wind power capacity (see tables 1 and 2). Responding to this market demand, Vestas has invested heavily in building new factories and assembly plants in those two countries. The massive scale of turbine equipment makes it important to keep transportation distances to a minimum. In addition to reducing costs, closer proximity cuts down on the greenhouse gas emissions associated with freight haulage—something this environmentally attuned company monitors closely.

Lone's English is excellent, but she chooses her words slowly and carefully. She is wary of outsiders wanting to know details about the company's closely guarded manufacturing processes.

As we set out on our tour, Lone tells me about the company's founder, a blacksmith named H.S. Hansen who began making farm equipment in the village in 1898. Those early days are commemorated by a quaint outdoor cluster of iron figures depicting workers in the old Lem Forge, just a few blocks from the village center, with its handful of small shops sprinkled among modest red-brick homes.

By the end of World War II, H.S. Hansen's son Peder had taken over the business, naming it VEstjyskSTålteknik A/S (fortuitously shortened to Vestas) and shifting the company's focus to the manufacture of farm trailers, ship engine intercoolers, and hydraulic cranes. In 1978, just when I migrated from college to Capitol Hill to promote renewable energy, Vestas built its first experimental wind turbines. After trying a few designs, the company's engineers decided on a three-blade turbine that went into production the following year. Eight years later, the company made wind energy systems its single focus, and it soon grew to be the world's top-ranked turbine manufacturer.

In 2004, Vestas acquired its prime Danish competitor, NEG Micon, and the consolidated company supplied 34 percent of the 8 gigawatts (8 billion watts) of new wind power capacity installed globally that year.[2] Over the next six years, worldwide demand for wind turbines skyrocketed, reaching 40 gigawatts by 2010. Vestas remained number one in the trade, but its market share dropped to about 12 percent of all new installations.[3] Still, the company brought in $9.2 billion in revenues that year and added nearly 3,000 people to its workforce—healthy signs of strength coming out of the global recession.[4]

Lone informs me that Vestas operates factories for blades, towers, and other wind turbine components at more than two dozen locations across Europe, Asia, and North America. Her job is to handle staff communications for the company's eight blade plants worldwide, fostering a sense of staff camaraderie across cultures and time zones from Colorado to China. Now and then, the company hosts planetary parties using satellite-beamed video-conference technology to connect workers celebrating at different plant locations around the world.

Rock music blares in the background as we step out onto the factory floor, our feet in company-issued red plastic clogs, our heads crowned by clean white hard hats, and our eyes shielded by safety glasses. This vast room, as brightly lit as the music is loud, is shaped like an enormous shoebox, nearly as long as a football field and about a third as wide. Lone first points to a machine that looks like an industrial-scale rotisserie. Workers clad in white protective clothing carefully feed long sheets of epoxy-saturated fiberglass around this squared-off steel mandrel, gradually expanding its girth. Layer upon layer of fiberglass is added, reinforced by wavy black veins of carbon fiber. Once heated, dried, and hardened, the resulting obelisk—more than a hundred feet long—will become the structural mainstay of a blade that is expected to withstand the forces of nature for two decades or more. In the trade, this structural spine is called the *spar*.

Lone directs me to the opposite end of the production room, where another team of workers is painting a protective sealant onto the inside surfaces of a giant double-mold that looks like an opened, elongated clamshell. Once the sealant has dried, epoxy-soaked fiberglass will be layered on each side of the mold, forming the shape of the blade's two halves. The bivalve will then be closed around the spar, an epoxy compound will be injected, and the parts will be fused into a completed blade.

The making of wind blades is a curious mix of high-tech engineering and traditional craftsmanship, demanding the sophistication and subtlety of aeronautic design along with the patient attention to detail found among builders of high-end yachts. When I ask how long it takes to produce a single blade, Lone is visibly uncomfortable. "That's the type of information our competitors want to know," she explains.[5] It's clear, though, that progress is measured in hours, even days. Producing 6,000 to 10,000 blades annually at plants around the world, the company's simultaneous demand for prodigious output and painstaking precision is daunting.[6]

A high-pitched, scratchy whine greets us on entering the factory's finishing department, where a team of masked workers uses handheld sanders to massage the creamy outer surfaces of each giant blade. Meticulously they shape the razorlike edges that will slice through the wind as efficiently and quietly as advanced aerodynamics allow. After all the surfaces are smoothed and painted, a tractor pulls each completed blade, mounted on a rubber-wheeled dolly, out into the factory yard. There a forklift gingerly stacks the blades, three and four high, in neatly ordered frames like bottled wine shelved in a vintner's cellar, waiting their turn to be shipped to customers.

Though Danish innovators built a few dozen power-generating wind turbines in the early twentieth century, the seeds of Denmark's modern wind industry were planted by the 1973 oil crisis. Denmark at the time was more than 90 percent dependent on oil for its energy needs, and virtually all that oil was imported. To punish Western Europe and the United States for their support of Israel during the Yom Kippur War, Arab oil producers radically reduced the flow of oil to those countries, causing long gas station lines and tripling the price of oil. Denmark's political leaders recognized the need for decisive action, but initially the government focused more on fuel-switching than on energy efficiency or renewable energy development. The nation's first comprehensive energy plan, released in 1976, emphasized coal, natural gas, and nuclear power as the primary pathways to reduced dependence on foreign oil. In the years that followed, North Sea explorations yielded substantial supplies of oil and natural gas—enough to make Denmark a substantial net exporter of both resources. No nuclear plants were built, however, and in 1985, responding to broad popular opposition, the Danish parliament officially banned their construction.

The same was not true for coal, which quickly gained a foothold as the primary fuel for the country's power plants. Like pre-embargo oil, nearly all of Denmark's coal was imported, but it came from a more reliable and diverse array of nations, including South Africa, Russia, Colombia, Australia, and the United States. Denmark's heavy reliance on coal continues to this day, despite an official ban on its use in new power plants as of 1997 and a commitment to phase out all coal-generated electricity by 2030. The pervasive reuse of waste heat from power production has substantially improved the overall efficiency of coal use in Denmark, with 60 percent of homes now kept warm during the long Scandinavian winters by a vast network of steam pipes running from the country's power plants.[7] Nevertheless, the energy sector remains the main contributor to Denmark's greenhouse gas emissions, and within that sector, coal is the overwhelming culprit.

Wind grew more gradually to its current 20 percent share of Denmark's power supply. The same year that the government released its 1976 energy plan, Danish academicians produced their own, emphasizing the need for a major shift toward renewable energy. Two years later, a test lab and certification facility for wind turbines was established at Risø National Laboratory, and the year after that, the Danish parliament adopted a new law that gave a big boost to wind turbine ownership: wind power producers of any scale would be entitled to an up-front government grant amounting to 30 percent of the purchase price of the turbines, for all models approved by the Risø laboratory.[8]

Making it easier to buy turbines was important, but two other steps were needed for wind energy to take hold. First, turbine owners had to be assured that the power generated by their wind machines would find its way onto Denmark's electric grid. With the government eager to encourage new wind projects, state-owned transmission companies agreed to cover 35 percent of the costs of connecting windmills to the grid. Second, turbine owners needed to be able to sell the power generated by their windmills at a reasonable profit. Here too, the major state-owned utilities settled on a formula that guaranteed wind power producers substantially more than utilities were paying for conventional coal-fired power.[9]

With these incentives in place, tens of thousands of Danish landowners decided to enter the wind business. Some planted single, grid-connected turbines on their properties. Others joined local cooperatives, or *vindmøllelaugene*, that pitched in together to build small arrays of turbines.[10] Local ownership of wind power was seen as a virtue, by the landowners themselves as well as their political supporters in Copenhagen. When nonlocal speculators rushed in to take advantage of the public subsidies, the government restricted its grants to people living within 10 kilometers (6.2 miles) of their windmills or in the same municipality.[11]

Denmark's utility leaders soon became nervous as they watched windmills sprout across the countryside. To protect their large, central-station power plants, they successfully lobbied for the removal of the investment grants to wind developers. In 1989, the grants were abolished,[12] but wind investors continued to enjoy guaranteed rates for the electricity they generated. They also benefited from a refund of the tax on carbon emissions that the Danish government charged to conventional electricity providers. With post-tax rates of return ranging from 5 to 22 percent per year, wind energy remained an attractive investment,[13] and the number of investors continued to grow. According to one account, 50,000 Danes held ownership stakes in windmills in 1991—nearly 10 percent of the country's population.[14] By 1999, 150,000 Danes were reported to be full or partial owners of wind turbines,[15] and during that period, more than 80 percent of the country's wind turbine capacity was locally owned.[16]

Appealing though it was for Danish landowners to invest in wind power, government planners quickly recognized the importance of building an export market for the country's emerging wind turbine industry. In a country with a smaller population than Massachusetts and little more than twice its physical size, it was clear that wind turbine producers would need to reach buyers outside Denmark's borders if they were to thrive. In 1982, the Ministry of Industry commissioned a report that looked at America as the next Danish wind energy frontier. The report triggered an immediate response among the country's leading turbine companies. According to one historian, sales representatives from Vestas and three other Danish turbine manufacturers were on a plane to the United States within a few days of the report's release, eager to test out their prospects.[17]

California was the prime target for Danish wind companies. A volatile combination of idealistic fervor and entrepreneurial bravado had spawned a new generation of California wind developers who were eager to cash in on some remarkably enticing state and federal investment incentives. As in Denmark, the 1973 oil embargo had sent shockwaves across America. Drivers long accustomed to cheap and plentiful gas suddenly found themselves lining up for hours outside filling stations. "Rationing," a term that hadn't crossed most Americans' lips since World War II, reentered the nation's vocabulary. And America's political leadership awakened to the folly of depending on imported oil to meet the nation's energy needs.

Richard Nixon, beleaguered by ongoing investigations into the Watergate scandal, grabbed hold of the energy crisis as a defining moment in his waning political career. On January 23, 1974, he delivered a Special Message to the Congress on the Energy Crisis, in which he outlined a lengthy slate of measures to bolster U.S. energy self-reliance. Along with creating a provisional framework for gas rationing, he called for expanded surface mining of U.S. coal, new mineral leases on federal lands, expanded offshore oil and gas production, and the accelerated licensing and construction of nuclear power plants. Research and development into "new" technologies was another feature of his program, though "solar electric power" was the only renewable technology mentioned, as a prospect "for the far term," at the end of a long list of more immediate priorities focusing on fossil and nuclear fuels.[18]

The historic nature of Nixon's energy agenda was made perfectly clear in his State of the Union address a week later: "In all of the 186 State of the Union messages delivered from this place," he said to the packed House chamber, "this is the first in which the one priority, the first priority, is energy." Under the rubric "Project Independence," he set a high bar for immediate progress in weaning America off of foreign oil: "Let this be our national goal," he said, barely glancing up from his text, with signature sweat beading on his chin and upper lip. "At the end of this decade, in the year 1980, the United States will not be dependent on any other country for the energy we need to provide our jobs, to heat our homes, and to keep our transportation moving."[19]

That goal was far from met. In 1973, 28 percent of our oil came from abroad: 6.3 million barrels a day out of a total U.S. daily consumption of 17.3 million barrels. Today we use substantially more oil than we did at the time of the embargo (19.1 million barrels a day in 2010), and we import about 60 percent of it.[20]

While the 1973 embargo and its aftershocks failed to lessen America's oil appetite over the long run, it did open the door to a new level of public and private commitment to renewable energy. In the early days of Jimmy Carter's one-term presidency, energy independence emerged as the dominant theme. Just two weeks after taking office in the winter of 1977, President Carter appeared before the U.S. public in a televised address that came to be known as the "sweater speech." Bundled in a beige cardigan and sitting by a glowing White House fireplace, the president called for a major commitment to energy conservation. In terms more befitting a pastor than a U.S. president, he decried Americans' wasteful ways and exhorted the public to "make modest sacrifices" and "live thriftily." He also signaled the need for new energy technologies like solar power, but that message was largely lost on American TV viewers, who came away feeling stunned by the president's call for a new age of austerity in which thermostats would be kept at a cool 65 degrees during the day and an even chillier 55 degrees at night.[21]

In his second televised energy address in April 1977, Carter struck a more conventional pose behind his desk in the Oval Office, but again he adopted a sermonizing tone. "Tonight," he opened, "I want to have an unpleasant talk with you about a problem unprecedented in our history." He went on to discuss the need to "balance our demand for energy with our rapidly shrinking resources," and he cautioned that his proposed policies would "test the character of the American people" and be "the 'moral equivalent of war,' except that we will be uniting our efforts to build and not destroy."

Jimmy Carter may not have won the hearts of the American public with his grim admonitions, but his administration's policies created a wholly new federal climate for advancing renewable energy. For Carter, as for Denmark's political leadership of the 1970s, coal was a big part of the solution to foreign oil dependence; he saw U.S. coal as "our most abundant energy resource" and advocated for its expanded use, along with the carefully supervised use of nuclear power.[22] At the same time, he saw a real role for alternative energy resources including the sun, in which he invested major symbolic value. In 1977, he called for solar energy to be used in more than 2.5 million American homes, and two years later he led by example, installing solar water-heating panels on the White House roof.

A further step that Carter took to advance solar energy was the creation of the Solar Energy Research Institute, a new national laboratory with Denis Hayes at its helm. Charismatic and articulate, Hayes had been the driving force behind the first Earth Day in 1970, and had more recently orchestrated the presidentially proclaimed Sun Day in May 1978. Hayes brought a broad definition of solar energy to the new national laboratory, including "anything that uses sunlight within a few decades of the time it arrived at the surface of the planet." In his view, this included wind because of the sun's crucial role in creating the thermal gradients that draw air at different speeds and in different directions across the planet.[23] As a practical matter, though, the research supported by his institute focused primarily on a narrower realm of solar thermal and solar electric technology.

While solar energy received greater fanfare, Carter's policies brought enormous benefits to the wind industry as well. Most consequential was the Public Utility Regulatory Policies Act of 1978 (PURPA), which gave small-scale power producers a way to compete in a market that had, for decades, been monopolized by large public utilities. Wind farms and other renewable energy-based power plants were PURPA's primary targets, along with small "cogeneration" facilities that made use of the waste heat created from power production. Public utilities were obligated to buy the electricity generated by independent generators at a price that was "just and reasonable" to consumers and, at the same time, did not discriminate against this new breed of power providers.[24]

PURPA predictably triggered a backlash by a number of public utilities that were reluctant to open up their markets to competition. Some states also resented the federal government's encroachment onto turf traditionally dominated by their own public utility commissions. Two court challenges to PURPA's reach into state-held territory went all the way to the U.S. Supreme Court, but the Court ruled in PURPA's favor in both cases. This cleared the way for a new generation of daring and often idealistic energy entrepreneurs.[25]

Just as PURPA guaranteed wind developers a market for their electricity, a powerful blend of tax incentives drew investors into the field. Under the Energy Tax Act of 1978, wind energy projects enjoyed a double benefit: they could claim a business investment tax credit of ten percent, and they were on a short list of special energy projects that qualified for a second ten-percent tax credit for capital outlays.[26] Then, in 1980, the Crude Oil Windfall Profits Tax Act increased the energy portion of the tax credit to 15 percent for solar, wind, and geothermal investments, bringing the total federal credit available to wind farmers to 25 percent.[27] What's more, commercial wind farms—like many other energy companies—benefited from federal tax provisions that let them depreciate their equipment on an accelerated basis.

If Jimmy Carter was the stern preacher waging a righteous war for American energy independence, California governor Jerry Brown was the smart and savvy iconoclast, surrounding himself with creative people who were willing to challenge conventional thinking on everything from religion and social relationships to the environment. Thirty-six years old when he became governor for the first time in 1975, Brown later recalled his early days in office: "This was the mid-1970s. It was the time of the Whole Earth Catalogue. I was dealing with people like Stewart Brand, Wendell Berry, Amory Lovins, Herman Kahn, and Dick Baker from the Zen Center. I mean, it was a hotbed of ideas. And there was a sense that we were on the threshold of a new politics. We were building something new. It was very exciting."[28]

An eclectic and contradictory mix of community empowerment and space-age zeal inspired Jerry Brown's politics, leading a famously caustic newspaper columnist, Mike Royko, to brand him "Governor Moonbeam." On one hand, Brown proposed that California launch its own communications satellite; on the other, he was an ardent proponent of backyard composters and rooftop solar panels. In the latter realm, he took particular inspiration from Amory Lovins, author of a widely read manifesto mapping out the social, economic, and political reasons to abandon our reliance on fossil fuels and centralized power generation.

In fact, Lovins's proposed paradigm for autonomous, community-based energy systems was much more radical than anything that Brown was able to advance as governor. Lovins was unequivocal in his condemnation of the status quo: "Siting big energy systems pits central authority against local autonomy in an increasingly divisive and wasteful form of centrifugal politics," he wrote in his 1977 book *Soft Energy Paths*. "In an electrical world, your lifeline comes not from an understandable neighborhood technology run by people you know who are at your own social level, but rather from an alien, remote, and perhaps humiliatingly uncontrollable technology run by a faraway, bureaucratized, technical elite who have probably never heard of you."[29]

Steps were certainly taken under Brown's leadership to help Californians recapture at least some of the power that they had drawn from distant, centralized generating plants for decades. A year after taking office, the governor turned to another energy mentor, Wilson Clark, and asked him to oversee the creation of California's Office of Appropriate Technology.[30] Under Clark's guidance, this new agency sponsored demonstration projects and disseminated information on small-scale energy systems, along with other residential and community-based projects. The Brown administration also introduced hugely enticing tax incentives for residential renewable energy investments. Californians could claim 50 percent of the price of qualifying equipment as a personal income tax credit, up to a total of $3,000 per household.[31]

While thousands of homeowners used the residential tax credits to purchase rooftop solar water-heating panels, the state adopted a model for wind energy that strayed far from Lovins's idealized vision of grid-liberated local self-reliance. With new wind farms selling their power to big utilities like Southern California Edison and Pacific Gas & Electric, the state's renewable energy marketplace would soon be drawn back into the "centrifugal politics" of big energy systems that Lovins railed so passionately against.

Various state agencies under Jerry Brown were instrumental in helping this process along. Surveys conducted by the California Energy Commission identified sites for 13,000 megawatts of wind capacity in the state; half of these were viewed as ripe for near-term commercial development. Particularly promising were a number of the mountain passes where cool air blowing in from the Pacific accelerated as it flowed into warmer inland regions. The Altamont Pass, east of the Bay Area, was one such area. Two others were the Tehachapi Pass, southeast of Bakersfield, and San Gorgonio Pass, near Palm Springs, both drawing swift and sustained air currents from the west into the Mojave Desert.

Along with charting the winds, California offered a 25 percent tax credit to commercial wind investors, and it gave tax-free bonding authority of up to $10 million to "alternative energy" projects.[32] California Democratic congressman Pete Stark minced no words in decrying these subsidies. "These aren't wind farms," he grumbled. "They're tax farms."[33] Added to the federal tax incentives, the state enticements were enough to create what many have likened to the Gold Rush that had swept the state more than a century earlier. Paul Gipe, an early publicist for the wind industry, estimated that California's wind developers raised $2 billion from as many as 50,000 individual investors during the boom years of the early 1980s.[34]

As wind investment capital flooded the California market, a scrappy collection of small, undercapitalized companies hustled to get their largely untested wares into the field. In 1981, wind developers at Tehachapi Pass installed 150 windmills; the following year they installed another 1,200 turbines; and in 1984, they topped out at 4,732 new units. By 1985, a grand total of 12,553 windmills were in place at California's three biggest wind energy complexes, and by 1987, more than 17,000 windmills had been installed statewide. The turbines came from more than a dozen manufacturers and ranged in size from a few tens of kilowatts up to 330 kilowatts of installed capacity per machine.[35]

While federal and state subsidies supercharged the wind farm investment climate, the government did little to make sure that turbines entering the market could actually perform as intended. There was no official rating system for turbines, and federal research-and-development (R&D) funding for wind technology singled out megawatt-plus prototypes—equivalent in scale to today's commercial turbines but several times larger than the machines that were entering the market in the late 1970s and early 1980s. Much of the federal research funding went through NASA, which regarded wind energy as a promising new focus for a flagging aerospace industry and awarded plum contracts to corporate giants such as Boeing, McDonnell Douglas, Grumman Aerospace, Westinghouse, and General Electric (GE).[36] According to one estimate, $350 million—almost three-quarters of total U.S. R&D spending on wind between 1974 and 1992—went to research on 1-, 2-, and 3-megawatt turbines, none of which made it to market.[37]

Fledgling U.S. wind manufacturers learned their hard lessons in the field, experimenting wildly with a dizzying array of designs. Some had airplane-style propellers mounted on a horizontal shaft; others rotated around a vertical axis, their slender, convex metal blades resembling giant eggbeaters. One brand of turbine failed after only a few days of operation; it had been advertised as lasting a minimum of ten years.[38] On other machines, gearboxes and generators failed, rotors spun out of control, blades cracked and collapsed, and towers buckled.[39] In their rush to build windmills that could produce power on a commercial scale, some manufacturers simply relied on scaled-up versions of small residential wind generators; these often-flimsy machines were simply no match for the turbulent winds of California's mountain passes. As Dutch technology analyst Rinie van Est commented about this first generation of U.S. wind turbine manufacturers, "they wanted to start dancing before they had learned to walk."[40]

If U.S. turbines were inspired by the aerospace industry and its lofty ambitions, Danish manufacturers were tied to a tradition of rugged and enduring craftsmanship, firmly grounded in traditional agriculture and industry. American wind developers were ready to recognize the difference, with one of them pointedly observing that rockets and missiles, which played out their useful lives in no more than an hour, were a poor precedent for wind generators that needed to run reliably for years, if not decades.[41] Just as Vestas could draw on decades of experience manufacturing farm trailers and cranes, another leading Danish company, Bonus, built irrigation systems, and a third, Nordtank, produced oil and water tanks.

The pragmatic spirit of these companies was reinforced by the Risø National Laboratory. Rather than focusing on speculative research into multi-megawatt prototypes, the lab devoted itself to testing and certifying much smaller market-ready designs. Its efforts converged with leading Danish manufacturers in favoring a single design that had demonstrated its reliability: a three-bladed rotor mounted on the upwind side of a supporting tower, with twin generators and a "yawing" system that mechanically adjusted the rotor to keep its blades facing into the wind. Experience gained operating turbines in Denmark strengthened the reputation of companies like Vestas as they pursued U.S. buyers. According to the Danish Wind Industry Association, some Vestas models had logged as many as 15,000 hours of operating time before being sold across the ocean. This compared very favorably to U.S. models, many of which found their way to California's wind farms after no more than 1,500 hours of operation.[42]

With American turbines failing at an alarming rate in the early 1980s, the promise of greater reliability generated an American boom for Danish wind technology. In 1983, Danish turbines accounted for 11 percent of new installations in California; the following year they occupied 33 percent of the market, and by 1987, they had reached a staggering 90 percent of all new units.[43] But even the relatively sturdy Danish machines had problems standing up to California's swift, blustery winds and dusty desert conditions, which placed much greater strain on turbines that had operated relatively well in Denmark's milder, steadier winds. Danish manufacturers were unfortunately slow to respond to these issues, in part due to a lack of timely information. Most California wind developers serviced their own equipment rather than relying on the original manufacturers, so company engineers back in Denmark had a hard time getting reliable feedback on issues as they arose.

While the 1980s were turbulent times for Danish as well as U.S. wind turbine models, policy changes at the state and federal level played an even bigger role in breaking the frenetic pace of California's wind farm construction in the latter part of the decade. Since 1978, wind developers had benefited from a federal law that made it all but impossible to build new power plants primarily fueled by natural gas or petroleum. The Powerplant and Industrial Fuel Use Act of 1978 barred the use of these fuels for most new electric generation, with the goal of stimulating an increased reliance on coal, nuclear, and, to a lesser extent, renewables—energy sources that were seen as less vulnerable to the fuel shortages that plagued U.S. industry and American consumers in the 1970s.[44] By 1986, expanded domestic gas exploration, together with reduced demand, caused a market surplus and a drop in gas prices that led Congress to lift restrictions on the fuel's use for power production.[45] Opening up the power market to natural gas made it much harder for wind and other renewable energy technologies to compete.

The other big problem facing wind developers and turbine manufacturers alike was the demise of both the federal and California tax credits that had been crucial to attracting wind farm investors. In the fall of 1985, Congress, with the full backing of the White House, refused to extend the federal investment tax credits for wind and certain other renewable technologies beyond December 31 of that year.[46] The following spring, President Ronald Reagan ordered the removal from the White House roof of the solar panels that Jimmy Carter had installed seven years earlier. One of these panels is now on exhibit at the Smithsonian, a bittersweet symbol of two presidents' very different visions of America's energy needs.

Ronald Reagan, even before taking office in January 1981, declared open war on Jimmy Carter's energy policies, which he linked to our "disintegrating economy" and decried as "based on the sharing of scarcity."[47] Digging out more coal, drilling for more oil and gas, and building new nuclear plants became the cornerstones of Reagan's energy vision; advancing renewable energy did not really fit with this plan. Once elected, he canceled the federal tax credits for renewable energy projects and slashed federal funding for renewable energy R&D. From a high of $718.5 million during Carter's last year in office, research dollars for renewable energy dropped to $110.8 million by the end of Reagan's two-term presidency. Wind's share of the R&D budget dropped from $77.5 million to $8.7 million during the same period.[48] The political momentum behind wind in Washington was gone.

In Sacramento, Republican governor George Deukmejian took office in 1983 with as much esteem for Jerry Brown's energy policies as Reagan had for Carter's. Stories about wind investment deals providing "welfare for the wealthy" and news footage showing broken-down wind turbines littering the California hills couldn't have helped. But even without the damning media reports, a statewide budget crisis gave Deukmejian the political cover he needed. In 1986, he rolled back the investment tax credit for commercial wind projects from 25 to 15 percent. The following year, he eliminated it entirely.

Even after the plug was pulled on the state and federal tax credits, wind farm developers continued to install new turbines, thanks largely to contracts that had already been signed for the long-term supply of wind power to California's two big utilities, Pacific Gas & Electric and Southern California Edison. As these contractual obligations were met, however, new installations began a steep and steady decline. In 1985, nearly 400 megawatts of new wind power came online. In 1986, that number dropped to 275 megawatts; from there it plunged to 154 megawatts in 1987. New installations hit an all-time low in 1992, with new turbines adding up to just 19 megawatts of new capacity.[49]

As new contracts for wind farms dried up, American and Danish turbine manufacturers alike fell into a deep slump. Their overreliance on a single market was taking its toll.[50] Vestas had increased its workforce from 200 to 870 employees in 1982, largely to fulfill orders from a single large California wind developer. When the tax credits evaporated and petroleum prices plummeted in 1986, it was left with a huge inventory of turbines that it couldn't sell. That year Vestas and a Danish blade manufacturer declared bankruptcy. The next year Nordtank followed suit, and in 1988, four other Danish wind energy companies declared their insolvency.[51] In the aftermath, a few firms, like Vestas, trimmed their staffs, reorganized, and survived; others disappeared.

The collapse of the California market could have driven Vestas out of the wind business. Instead, the company brought in new leadership and devoted itself to building new customer bases across Europe, Asia, the Pacific Rim, and other parts of the United States. In 1988, it won a bid to build six wind farms in India, financed by the Danish International Development Agency. A year later, it opened a subsidiary in Germany, and in the early 1990s, it expanded into new markets in Sweden, the United Kingdom, Australia, and New Zealand. In 1994, a joint venture with a local turbine producer, Gamesa, brought Vestas to Spain.[52]

Germany and Spain proved to be great strategic bets for Vestas and other wind energy entrepreneurs in the 1990s. In Germany, a law enacted in 1991 obligated utilities to pay solar and wind generators a fixed "feed-in" tariff amounting to 90 percent of the average consumer price for power. Along with enjoying a guaranteed price for their electricity, wind developers benefited from low-interest loans as well as planning guidelines that made it easier to site new projects. These incentives led to a dramatic surge in German wind power, from only 31 megawatts in 1990 to more than 6,000 megawatts of installed capacity a decade later—nearly a 19,000 percent increase! This made Germany number one in wind power generation worldwide.

Spain experienced similarly dramatic growth during the 1990s. From only 4 megawatts installed as of 1990, the country boosted its wind power capacity to 2,836 megawatts by the year 2000, placing it second only to Germany. As in Germany, this dramatic growth was triggered by a tariff that guaranteed Spanish wind producers a highly favorable rate for their electricity. Producers were paid the market price for power plus a premium that, in 2000, amounted to three euro cents per kilowatt hour.[53]

During the 1990s, Denmark continued to ramp up its own use of wind, stimulated by a succession of pro-wind government plans as well as a feed-in tariff that guaranteed wind producers an above-market rate for their electricity. This was also the decade when Denmark opened up a vast new horizon for renewable energy development: offshore wind farms. In 1991 and 1995, two pilot projects tested out the technology, with results so encouraging that the government, in its 1996 energy plan, set a target of building 4,000 megawatts of offshore wind power by 2030. By the year 2000, Danish wind power—still mainly on land—totaled 2,291 megawatts, ranking it fourth in the world, just behind the United States, which reached 2,539 megawatts of installed capacity that year.[54]

The last leg of my Vestas tour in Denmark took me to the company's global headquarters, an attractive modern building a short distance from the city of Aarhus. There I met Peter Wenzel Kruse, the company's communications chief. Short in stature with a wiry build, he spoke in staccato phrases about his company's global positioning. He sees Denmark as a showcase for wind power, and in that spirit he worked hard to land Vestas a high-profile spot at the Copenhagen climate summit in December 2009. Visiting VIPs and reporters who attended the event could hardly miss the prominently labeled Vestas turbine that towered above the meeting's venue, a low-slung conference center on the outskirts of the capital city. Kruse said the exhibited turbine was "a good gimmick," but he quickly added: "We have no business in Denmark."[55]

Denmark may be ahead of all other nations in the *percentage* of its power generated by wind. But in the *amount* of electricity from wind, it barely earns a top-ten global ranking, lagging behind six other European nations plus India, the United States, and China (see tables 2 and 3). Even with the Danish government's plan to provide most of the country's power from wind, Kruse stresses that the local market for turbines will be minuscule relative to overall global demand. Denmark, after all, is a country with 5.5 million people in a world whose numbers already exceed 7 billion.[56]

As one sure sign of the company's shift to global production and marketing, Vestas shut down four of its ten factories in Denmark in 2010, and laid off 2,000 Danish workers, bringing the company's total domestic labor force down to 5,600. The Lem blades plant survived, but layoffs hit that factory, too. In October of that year, Vestas CEO Ditlev Engel announced that the company would expand its U.S. workforce from 2,300 to 4,000 during 2011, filling slots in freshly built blade, tower, and nacelle factories in Colorado. By mid-2011, the company's American workforce had reached 3,100 and was expected to continue growing.[57]

In all, wind turbine production has created about 20,000 new American factory jobs, and this number is expected to grow as turbine companies and their suppliers gear up to meet America's appetite for wind power.[58] Wind power in America may have hit the doldrums following the California feeding frenzy of the 1980s, but its recent reemergence as a world wind energy leader has attracted vigorous competition among eager technology suppliers. Vestas is far from alone among foreign corporations that are seeking out U.S. buyers for their wind machines. Some of these businesses—like Spain's Gamesa and

India's Suzlon—are hardly known outside the wind trade. Others are familiar global conglomerates like Siemens and Mitsubishi (see table 4). Whatever their corporate roots, most turbines sold in America are assembled on U.S. soil, with more than half of their components and subcomponents domestically made. Exploring this growing U.S. manufacturing base would be the next stop in my travels.

CHAPTER THREE

Rust Belt Renewables

YOU DON'T HAVE TO BE an industrial giant to be a player in the American wind business. American Superconductor, a Massachusetts firm, markets and licenses state-of-the-art turbine designs to global manufacturers. In the Great Lakes region, about seventy firms are involved in manufacturing turbine components and subcomponents. One of them, just outside Cleveland, is a family-owned bolt-forging company called Cardinal Fastener. When Barack Obama made his pre-inaugural whistle-stop tour from Chicago to the nation's capital in January 2009, he made a point of visiting the Cardinal Fastener forge, where he talked up wind energy's importance in a spirited meeting with plant workers. In a video clip documenting the event, two muscle-bound workers firmly grip a yard-long piece of gunmetal-gray hardware. "This is a 2 ¼-inch heavy hex bolt," one of them says as he peers at the camera through his protective eyewear. "Now it's called the 'Obama Bolt.'"[1]

General Electric supplies nearly half of America's wind turbines, but most of the equipment in its machines comes from smaller manufacturers scattered across the country and around the globe. Though GE has no real competition among U.S. turbine producers, one much smaller company, Clipper Wind, has worked valiantly to establish itself in the market. Clipper's road has been a rocky one, fraught with technology flaws and financial distress, but it has brought new jobs and new hope to hundreds of workers at its turbine assembly plant in Cedar Rapids, Iowa.

Small and sparsely populated, Iowa has captured a surprising number of new wind energy manufacturing jobs. Ranking thirtieth in the nation in population, the Hawkeye State has the fourth-largest workforce commitment to wind manufacturing, outflanked only by Texas, Illinois, and Colorado.[2] When fully staffed, nine wind-dedicated factories scattered across cities and towns in southeastern Iowa will employ 2,300 workers.[3] Some of these jobs are in foreign-owned workplaces like the blade factory that Siemens built in Fort Madison and the turbine assembly plant that Acciona, a Spanish company, now operates in West Branch. Others are in American-owned factories like the Trinity Structural Towers plant in Newton, Iowa, a major supplier of turbine towers to GE. Across a stretch of cornfields from the tower plant is a blade factory run by TPI Composites, a Rhode Island–based firm. These two operations have filled at least part of the huge employment gap created when Maytag shut down its headquarters and appliance assembly plant in Newton in 2007. And then there are the many Iowa firms—more than 200 of them[4]—that are part of a supply chain providing the 8,000 components and subcomponents that make up the modern-day wind turbine.[5]

Clipper began assembling wind turbines in Cedar Rapids in 2006, and it has since brought approximately 350 jobs to a city that has seen one Rust Belt factory after another close down or move out of state. The company is a relatively small player, supplying 6 percent of the U.S. market in 2009 and only 1.4 percent in 2010, but it is emblematic of the hundreds of firms that are now contributing to the U.S. wind industry.

When I traveled to Cedar Rapids in February 2010, Iowa's premier industrial city still looked badly beaten by the flood that had engulfed it two years earlier. The Cedar River's roiling waters had driven more than 8,000 people from their homes and had forced hundreds of city businesses to close, some forever. On my visit, I found entire streets of modest clapboard homes nailed shut, empty of life. If it weren't for the snow on the unshoveled sidewalks and pitched rooftops, this could have been post-Katrina New Orleans. A few doors down from the boarded-up Paramount Theatre, I read a spray-painted message on an abandoned store window: "Bent, not broken."

Sited on the southern edge of the city, Clipper narrowly escaped the river's wrath. Veteran machinist Mark Meader told me the factory survived because it sits "on the highlands," actually only a few feet above the waters that flooded out Clipper's nearest downhill neighbor, Casey's General Store. Just as Clipper managed to dodge the flood, Mark and his coworkers Mick Boots and Dave Wheatley have been spared the worst impacts of heavy industry's flight from Cedar Rapids. All three men, in their early sixties, were hired by Clipper within the past few years. They consider themselves lucky to have found jobs in their fields.

Wind energy may be a rapidly evolving twenty-first-century technology, but it has strong roots in heavy industry, in the kinds of trades that Mark, Mick, and Dave have plied for decades. All three worked many years at Goss Graphic Systems, manufacturing newspaper printing presses until that factory closed in 1999. Mark worked in the machine shop, Mick in assembly, and Dave in engineering. "As a group, we were pretty much used to dealing with heavier, bigger, bulky items," Dave recalls. "It's nice to see this kind of industry come back into the area."

Bob Loyd is the person responsible for bringing his former coworkers at Goss over to Clipper. He ran the assembly operations at Goss, and he's now the plant manager at Clipper. "When we started off this plant, I literally picked up the phonebook and started calling people I knew," Bob recalls. "I knew good mechanics and electricians, guys who know gears. If you think about a printing press, it's a big gearbox with hundreds of gears." He realized that wind turbine machinery isn't so different.[6]

As he set about staffing the Clipper factory, Bob knew what he was looking for. He wanted a crew of seasoned industrial hands who could train a younger cohort coming from their first jobs or from courses in electronics, automobile mechanics, and engineering at Kirkwood Community College, just a mile or two up the road. Then, after five to ten years, the older guys could retire, having secured a next-generation workforce for this struggling Rust Belt city.

Born and raised in Wisconsin, Bob comes from a long line of mechanical and civil engineers. His great-grandfather had a bridge-building business; both grandfathers were engineers; and his father was the lead engineer at a company that built large diesel engines for U.S. Navy ships. College studies brought Bob to Iowa, where he started out at Iowa State and then transferred to the University of Iowa, earning an MBA and two engineering degrees. Despite his university credentials, he describes himself proudly as "a wrench turner."

"We all have dirt under our fingernails," Bob tells me as he leans back from the piles of paper on his desk. His broad hands rest firmly on the molded plastic arms of his chair, and he smiles. "I work on tractors, motorcycles, cars." His spartan office is just footsteps away from the vast factory floor, where workers assemble gearboxes, burnish cast-iron turbine hubs, and mount all manner of electronic equipment into the nacelles of Clipper's 2.5-megawatt Liberty turbine. The company is now developing larger machines for onshore and offshore wind farms, but the Liberty is Clipper's flagship.

The Liberty's Cedar Rapids birthplace, at 4601 Bowling Street SW, is a mammoth steel-frame shed whose history tracks the course of Iowa's rocky romance with heavy industry. It also holds important memories for Bob Loyd. The building opened in the mid-1960s, when Link-Belt Construction Equipment Company began manufacturing cranes and excavators at the site. Bob joined the company fresh out of college in 1973. By the mid-1980s, Link-Belt would be gone, bought out and moved south by a Japanese conglomerate, Sumitomo Heavy Industries. Bob later found himself back in the space as manager of the Goss printing press assembly plant. That company, too, did not last long in Cedar Rapids. Succumbing to fierce price wars with foreign competitors, Goss laid off hundreds of workers and eventually filed for bankruptcy.

Next in was Maytag, which used the building as a warehouse for refrigerators. By 2006, Whirlpool had acquired the fading Maytag brand and soon announced that it would be shutting down the company's headquarters and most of its assembly operations in Iowa. About 1,800 local Maytag employees, from office workers to factory laborers, lost their jobs.[7] Although some limited Whirlpool manufacturing remained in Iowa, there was no longer a need for warehousing at the Bowling Street facility.

When Bob was hired by Clipper, he welcomed the chance to breathe some new life into the battle-scarred factory space that he knew so well, and he was even more excited to help a few hundred people find jobs once again in heavy industry in Iowa. With all the local factory closures, he knew plenty of people who had been forced out of the field. Dave Wheatley was one of them. A skilled mechanical engineer, Dave drove a FedEx truck, worked as a taxi driver, and built windows after the Goss debacle. Mark Meader found a job as a carpenter. Others migrated to food processing plants around town. Many of these dedicated industrial hands, Bob knew, would jump at the chance to return to building some real machines.

For Bob, Clipper's allure went further. He had worked for Rockwell and Raytheon, so he knew that "heavy industry" often involves a heavy dose of defense contracting. "I wouldn't have been excited if we were making armaments for the military," he acknowledges. But when he talks about wind turbines and renewable energy, his voice becomes animated and his face brightens. "This is a product the country needs . . . and it's a great thing for the state. It helps the farmers, and it helps kids get good jobs."

Now nearing sixty, Bob has poured heart and soul into bringing youth and a hopeful future to the Clipper factory on Bowling Street. With obvious pride, he introduces me to four of the company's younger crew members. After seating us around a white laminate lunch table in the staff lounge, just off the factory supply room, he goes back out onto the factory floor, leaving us to talk. It's 3:30 in the afternoon and the room echoes with the pounding of the punch clock a few feet away. First-shift workers, mostly in their twenties and thirties, are streaming out of the building, and the second shift is coming in.

Matt Lalley is one of Bob's next-generation hires. Boyish-looking and a little shy, he tells me about his work as second-in-command in the electrical department, where he troubleshoots issues that come up in the assembly of generators and turbine control systems. His work as a car mechanic helped prepare him for the job, and what he didn't learn in the car repair shop, he has picked up on the factory floor.

Joel Peyton, twenty-seven, talks excitedly about his work as lead technician in the Remote Monitoring and Diagnostic Center, which tracks the performance of hundreds of Clipper turbines at wind farms stretching across the continent. He came to Clipper fresh out of college, where he studied management after training to be an electrician. Every day, he commutes forty miles each way from a farm where he and his father grow corn and beans. "It's our hobby," he says of the farming, adding that his father's real job is in Clipper's gearbox assembly division. A concrete worker for many years, he followed his son to the company just a few months after Joel came on board in October 2006.

Matt chimes in that *his* dad works alongside Joel's dad in gearbox assembly. Then Mary Tiedeman tells me that she has an uncle and a cousin who work in the plant. Mary, also very young, is one of the few women I have seen. Trained in child care and family services, she works as a receptionist.

"For the record, my dad *doesn't* work here!" the fourth in our gathering pipes up. Everyone around the table breaks into laughter. Tyler Glass, a twenty-one-year-old fix-it guy and 3-D computer designer, has just finished his training in mechanical engineering at Kirkwood Community College. Cherubic and animated, Tyler verges on euphoria when he talks about his job. "I love it—I love coming to work every day. It might sound cheesy, but I wouldn't say I'm here just to get a paycheck," he says. "I truly do believe in what we are doing."

Bob leads me through the gearbox assembly bay to a production zone where workers are busy installing electronic, mechanical, and hydraulic controls into Liberty nacelles. His people skills are obvious. Everyone gets a jovial, first-name greeting, followed by a question that shows Bob's awareness of exactly what that person is contributing to the Clipper effort.

The Clipper day begins at 7 a.m. with a shiftwide calisthenics class. In addition to staying fit, Bob wants people to feel part of the team. Some days he leads the exercises, but often he recruits others. "We always pick on somebody," he explains in a gentle tone that makes it clear this is an act of inclusion, not punishment.

Reflecting more broadly on workplace morale, Bob describes a quick-turnaround electronic suggestion box that invites workers to air their concerns. He calls it "VOICE," which is much easier to swallow than "Valued Operational Improvement for Clipper Excellence." Each month, the worker with the best idea gets a gift certificate.

Operational and design improvements are the name of the game for a small turbine company that is fighting its way into a market dominated by global giants like GE and Vestas. Technology glitches can be devastatingly costly, as the Clipper management team has learned in its bumpy start-up years. Joel Peyton is on the front lines in spotting problems as they pop up on the big-screen computer monitors that line the walls of the Remote Diagnostic and Monitoring Center, a crowded cluster of small rooms just around the corner from Clipper's staff lounge. We enter the center through a door prominently marked "Tornado Safe." In the stormy Midwest, I imagine this is no joke.

For every Clipper turbine across the continent, dozens of parameters are transmitted on a real-time basis to the center's computer data bank. Joel leans down, types in a few codes, and opens a window that displays data about the Fowler Ridge Wind Farm in Benton County, Indiana. Ambient and equipment temperature, wind and rotor speed, generator revolutions per minute (RPM), power output, and much more are monitored in real time for each of the forty Clipper turbines operating at this site.

A lot can go wrong when accelerating the rotational speed of a Liberty turbine from the top hub speed, 15.5 RPM, to the maximum generator speed of 1,133 RPM. Lubricating oil needs to be warm enough to flow to the spinning gears—a challenge in the super-cold weather that strikes many of America's premier wind sites. Pre-heating the oil is essential. "We don't want to push very thick oil through small holes," Joel explains. Without proper lubrication, bearings and gearboxes can overheat, causing a red icon to flash on the computer monitor and triggering an immediate shutdown.

Worker safety is another factor carefully watched by Clipper's thirteen-member remote monitoring and diagnostic team. Joel switches to a screen showing a satellite map of the continental United

States and beyond. Two circles surround every Clipper-equipped wind farm, a red one at a thirty-mile radius and a yellow one at fifty miles. Looking at the Bahamas, I point to brightly colored splotches on the screen, which Joel explains are lightning storms. "In the summertime, we can get pretty busy with lightning alerts," he says. "When technicians are busy up-tower, we want to give them a pre-warning so nobody gets hurt."

The Remote Monitoring and Diagnostic Center can make systemwide changes, where needed, to improve turbine operations. "We can adjust parameters here on 400 turbines and make sure it's done correctly," Joel explains. "Less human error, I guess you could say."

Even with this fine-tuning in the field, quality control poses a particular challenge for Clipper, as it does for several other turbine companies that purchase most of the components and subcomponents in their machines from outside vendors rather than manufacturing them in-house. Unlike Vestas, which produces much of its own hardware, almost everything that goes into a Clipper turbine is brought in from outside. Its generators come from Mexico. Castings for gears and hubs are supplied by vendors in the United States, Germany, Brazil, and Spain. Main shafts connecting turbine rotors to generators are forged in the United States and Slovenia. Blades are made by the same Brazilian firm that produces blades for GE's turbines. And towers come from as close as Chattanooga, Tennessee, and as far away as China. In all, Clipper relies on 120 outside suppliers for its turbine components and subcomponents.[8]

With parts coming from so many sources, maintaining uniformly high production standards demands constant vigilance. The steel used in gears, shafts, and bearings must be extremely durable to withstand the rigors of operating a machine that is in nearly perpetual motion, out in the elements, from one season to the next, year after year. Blades must be strong enough structurally and yet sufficiently flexible to accommodate widely varying wind conditions, and their surfaces have to withstand every kind of weather, from sub-zero cold to searing heat. And all moving parts must conform to very precise dimensional tolerances, ensuring their own internal integrity as well as their compatibility with components coming from other vendors.

To manage the global supply chain, Clipper has a full crew of quality-control technicians who inspect vendor factories across North and South America, Europe, and Asia. Quality-control experts are also on-site in Cedar Rapids, monitoring equipment as it arrives at the Bowling Street plant and testing key components on carefully calibrated machines. On our walk through the plant, Bob Loyd shows me a sound-insulated chamber, at the end of one long assembly bay, where every gearbox is run at up to 130 percent of maximum load under a variety of simulated wind conditions over a four-hour period. He likens this regimen to Detroit's factory testing of automobile drivetrains: both verify good mechanical performance but do not really mimic the stresses placed on equipment over many years of real-world operations.

For all its vigilance, Clipper encountered some serious defects in its Liberty turbine, starting in 2007. That year, the company reported a "supplier quality deficiency" that compromised the Liberty's drivetrain and gearboxes. Later the company divulged that more than a third of the turbines produced in 2006 and 2007 required drivetrain repairs. Defective blades added to Clipper's woes, requiring the blades on about 260 rotors to be reinforced. Most of these rotors had already been delivered to widely scattered wind farms, making repairs much more difficult and expensive. The combined price tag for repairs topped $107 million by the end of 2007.[9]

Cracked blades and other defects continued to plague the company in the years that followed. Repair costs escalated to $222 million in 2008, and it took until early 2010 to repair all the damaged blades.[10]

Clipper did what it could to minimize the costs of repairing its turbines in the field. Initially, to take down or put up a blade, Clipper repair crews would use a huge crane that had to be hauled to each site on fifteen to eighteen semitrailers. This cost the company a profit-killing $200,000 per job, or twice the price of the blade itself. But then the company's engineers came up with a new way to swap blades, using a technique that required only a few specialized tools and a cherry picker like the ones that phone company line crews use for repairs. The cost: $15,000 plus a few hours' labor. Similar savings were achieved by using the Clipper's built-in hoist, rather than a rented crane, to switch out faulty gearbox components.

Despite its attempts to streamline repairs, the burden of so much unscheduled maintenance took a severe toll on Clipper's balance sheet. Then the financial crisis hit, causing wind farm developers to slow their implementation of existing projects and freeze the development of many future ones. By September 2009, Clipper's cash reserves had dropped to a precarious $40 million. The company was in need of a bailout, and United Technologies Corporation came to the rescue.

Under the deal with UTC, Clipper initially surrendered 49.5 percent ownership to the Connecticut engineering giant in exchange for a pledge of a quarter-of-a-billion dollars. Ultimately, as Clipper's financial condition worsened, UTC ended up buying the company's remaining shares for $112 million in the fall of 2010.[11]

In taking on this relatively small, ailing enterprise, UTC's leadership saw a bigger strategic opportunity. Ari Bousbib, executive vice president and president of commercial companies at UTC, looked enthusiastically at the global market trends for the $50 billion wind industry. He commented to the press: "Other than maybe elevators in China, I don't know many industries that have grown at 25 percent a year."[12]

While acknowledging that his company urgently needed an outside infusion of capital, Clipper's chief commercial officer Bob Gates sees the UTC buy-in as much more than a financial bailout. He talks about the knowledge transfer the merger brings, given UTC's long history of manufacturing precision-engineered products ranging from Otis elevators and Sikorsky helicopters to Pratt & Whitney jet engines and industrial turbines. "You know, the elevators and escalators that Otis makes are all electronically controlled, and the elevator always stops at the right floor," he comments wryly. Gates is confident that Clipper will benefit from a higher level of quality control on the equipment it installs in its wind turbines.[13]

Just as Clipper is a story of Rust Belt renewal and pioneering innovation, it is a cautionary tale with some important lessons to companies that are looking to come in on the ground floor of an exciting but demanding new industry. Perhaps most obvious is the need for extreme vigilance before dispatching turbine technology of today's mammoth scale to far-flung wind farms. When blades get to be 150 feet long and are bolted onto hubs that are more than 250 feet in the air, companies better make sure they're sending out equipment that their customers can count on. The same is true for high-stress components like gearboxes and generators. Neither Clipper's quality-control teams nor the company's in-house simulators were successful at intercepting serious problems that later rang up such huge replacement costs in the field.

Another lesson relates to scale—not of the turbines themselves, but of the companies manufacturing them. Competing in the global marketplace for wind technology demands a level of financial commitment and an ability to absorb setbacks that are far beyond the means of relatively small, modestly capitalized companies like Clipper. Remember the Zilkhas' decision to bring in Goldman Sachs when Meridian Way and other wind farm developments needed a big-time infusion of capital; as savvy energy entrepreneurs, they knew when it was time to reach beyond their own resources. Clipper seemed to wait until disaster was at the door before it turned to UTC for a bailout.

While a company of Clipper's size simply may not have been equal to the challenges of fielding today's turbine technology prior to the UTC buyout, there *is* a place for smaller companies in the wind energy supply chain. With thousands of components and subcomponents going into the typical commercial turbine, the opportunities abound for specialized manufacturers to enter the industry. From manufacturers of bolts and flanges to the makers of hydraulic pumps and electric motors, companies of many shapes and sizes have already joined the American wind industry. These businesses stand to flourish as turbine companies at home and abroad—hopefully including a revitalized Clipper—expand their production to meet the growing demand for wind power in the coming years.

One of the conditions that will help the U.S. wind industry build and maintain momentum is a stable federal policy climate that puts renewable energy on an equal footing with traditional fossil fuel and nuclear technologies. Unfortunately, federal policies have been erratic and unreliable, inviting a degree of caution among energy investors, technology manufacturers, and wind developers that has impeded their full commitment to the enterprise.

Congress first approved a production tax credit for wind energy as part of the Energy Policy Act of 1992, with Senator Chuck Grassley of Iowa, fittingly, as its prime sponsor. Providing an income tax credit of 1.5 cents per kilowatt hour of electricity produced by wind, this law ushered in a wholly new approach to federal intervention in support of renewable energy.[14] Unlike the investment tax credits that gave commercial wind its start in the 1980s, the production tax credit—or "PTC" as it's called in the trade—rewards the actual generation of power rather than the mere building of new power-generating capacity. It gives wind farm owners and operators an incentive to operate their facilities as efficiently and productively as possible, minimizing downtime for repairs and maximizing the power produced by every operating turbine.

Though a valuable stimulant in concept, the PTC's actual impact has been compromised by the ebb and flow of legislative support over the years. Congress first let the PTC expire in June 1999, when its initial authorization under the 1992 energy law ran out. It was reinstated six months later, but for only two years. Then, at the end of 2001, it lapsed again for a number of months, and once more at the end of 2003 when a comprehensive energy bill authorizing its extension failed to pass Congress. When the PTC resumed in late 2004, its approved duration was little more than a year. And so the pattern continued up until the American Recovery and Reinvestment Act of 2009 extended the tax credit for a somewhat less truncated three-year period, through the end of 2012.

By the time this Obama-era legislation became law, inflation indexing had raised the PTC for wind-generated power to 2.1 cents per kilowatt hour.[15] What's more, wind developers could choose between claiming the per-kilowatt-hour credit and taking a straight-up 30 percent investment tax credit on qualifying equipment and expenditures. As a third option, they could convert the investment tax credit to an outright grant amounting to 30 percent of the value of their installations.[16] While some congressional lawmakers have attacked the degree to which these stimulus dollars have benefited foreign wind developers and manufacturers,[17] American wind industry leaders have credited the stimulus package with preventing a U.S. jobs meltdown. According to the American Wind Energy Association, the wind industry lost about 10,000 jobs when new turbine orders lagged and new wind farm construction slowed in 2010 to about half the pace of the previous year. This brought the total wind industry workforce down from a pre-recession peak of 85,000.[18] But with wind development regaining momentum in 2011, wind industry leaders are hopeful that jobs will regain and eventually outstrip their prior numbers.

Even with the infusion of stimulus funds, wind farm developers and equipment manufacturers remain wary of the future. Clipper's Bob Gates points to his own experience negotiating with vendors. "When you go to a foundry in Ohio and you say, 'Build a new plant to make more castings for wind turbine gearboxes,' the casting company says, 'What are you talking about? The PTC is going to expire in a year or two! I need ten or fifteen years to recover the cost of a plant.'"

"The PTC was meant to level the playing field economically, which it does," Gates continues. "What it doesn't do is level the playing field in the time dimension." He points to the uninterrupted subsidies enjoyed over decades by the oil, gas, and coal industries and wonders aloud why wind—so much more benign—has been supported so much more sporadically. As one sign of our nation's misaligned priorities, he laments our failure to internalize the huge environmental costs of coal: "You burn it—it's gone. You have the emissions, you have the CO_2, the energy source is gone, and for that, you get an incentive." I find myself mentally adding to the list: the blackened lungs of miners, the ravaged mountainscapes and polluted streams of Appalachia, the endless procession of railroad cars carrying coal from the surface-mined moonscape of eastern Wyoming to power plants across the country.

As for nuclear, Gates points to the indemnity that Congress granted to power plant operators back at the dawn of the civilian nuclear industry in 1957. Under the continuously reauthorized Price Anderson Act, even the most catastrophic meltdown would expose the plant operator to less than half a billion dollars in liability.[19] This level of federal protection for a civilian industry is without parallel in the United States, and its significance becomes very real when one contemplates the tens of billions of dollars of damage to property and people's lives caused by the recent catastrophe at Japan's Fukushima Daiichi nuclear plant.

"If a private company had to buy private insurance on a nuclear power plant, you wouldn't have any nuclear power plants," Gates flatly asserts. To that vast subsidy, we must add the billions of dollars the federal government has invested in trying to come up with a safe, long-term method for disposing of nuclear waste. Although no solution is in sight, civilian nuclear plants in America—104 in all—add to the nuclear waste burden every day. "How much does nuclear really cost?" He poses the question in exasperation, clearly not expecting an answer.

Building a U.S. wind manufacturing base extends far beyond the dozen or so companies that actually assemble turbines. The Timken Company, for example, is a successful industrial innovator that has long served the automotive and aeronautics industries. Today it is making a strategic bet on wind. The company's politically conservative leadership defies any stereotypes about the wind industry being led by progressively minded twenty-first-century entrepreneurs who are eager to help America move away from carbon-based energy technology and are gung-ho about the battle against global warming.

Headquartered in Canton, Ohio, Timken is a world leader in the manufacture of precision bearings, ensuring the smooth operation of everything from helicopter rotors and jet engines to car transmissions and off-road construction equipment.[20] With factories operating in twenty-seven countries, its 2010 sales totaled $4.1 billion.

Lorrie Crum, manager of Timken's global media relations and strategic communications, is passionate about the company's role in improving wind turbine performance. "Wind energy has been identified as the company's most promising market," she says, pointing to the $200 million that Timken has already sunk into pursuing its potential in this field. The fit is perfect, she says, for a company that prides itself on building for endurance—something of a mismatch with the automotive sector, where Timken's products substantially outlast the machinery that hosts them. "We were able to make the million-mile axle, but you never capture their full value," she says. All of those million-mile axles end up in the landfill, junked along with the rusting carcasses of cars that seldom travel more than a few hundred thousand miles.

Wind turbines, by contrast, need to operate day-in, day-out, season-to-season, year after year, over an expected lifespan of two decades or more. "This is extreme engineering," Crum explains.[21] Timken specializes in making bearings for the wind industry—meticulously crafted clusters of hardened steel rollers that hold key turbine components in place while allowing them to rotate with minimum friction. Ever since the company's founder, Henry Timken, began exploring ways to make sturdier wheel assemblies for horse-drawn carriages in the late nineteenth century, the study of friction, or "tribology," has been a driving force behind Timken's technology development.

One of the "mission-critical" bearings that Timken provides to wind manufacturers is a tapered bearing that cradles the main shaft of the turbine, a multiton steel rod that carries the slow-spinning motion of the rotor—usually in the range of 12 to 20 RPM—to a gearbox that speeds up to match the demands of the turbine's electrical generator. This bearing has to hold the main shaft firmly in place through all kinds of weather, allowing it to absorb the relentless jostling and wildly varying velocities of the winds in locations that are specially chosen because of their high prevailing wind speeds.

The damaged main shaft bearing from a large commercial wind turbine arrives just as we are touring the factory floor at the Timken Technology Center. Timken didn't make the original bearing. Rather, the Technology Center's engineers are being called upon to diagnose why it failed and to replace it with a more durable, custom-designed Timken bearing.

I watch as a diesel-powered mobile crane gingerly carries the main shaft assembly in through an open bay. Suspended by two triangulated wires from the crane's extended arm is a shiny silver-colored shaft, about ten feet long. At one end is the steel casing, about three feet in diameter, which contains the ailing main shaft bearing. Jim Charmley, who heads up the Technology Center's staff of 400 technicians, is on hand with me to witness this arrival. "We did *not* stage this for you!" he insists.

The main shaft bearing is designed to last the full twenty-year predicted lifetime of a wind turbine, but Charmley tells me that this one made it through only about two years. When I ask if many other main shaft bearings have failed on this particular turbine model, the MIT-trained mechanical engineer is circumspect. "I can only say that if [the manufacturer] had worked with us from the beginning, you might not see this here today." With that, he introduces me to Gary, a colleague whom he describes as one of the world's experts on nanocrystalline deposits—specialized metals like tungsten that can be applied to bearing surfaces to minimize friction and enhance durability. Gary, he says, will be incorporating nanotechnology into the bearing's redesign.

Charmley points to his own staff's dramatically shifted focus as a clear sign of Timken's commitment to wind. In 2005, there might have been three technicians working on wind applications at the Timken Technology Center. Five years later, he estimates that eighty to a hundred specialists, including many of the fifty PhDs working at the site, have been redeployed to develop new Timken products and services for the wind industry. He shouts out these impressive numbers as we walk through a cavernous room filled with forty high-speed electric motors that are testing out newly designed bearings. Salt and dirt are sprayed onto some bearings to test the durability of the rubber seals surrounding them; other bearings are jostled and shaken to test their ability to withstand tough conditions in the field. All these simulations are run on an accelerated schedule so that Timken can keep a leg up on the competition. Established European competitors like SKF and FAG are busily developing similar products, and Asian companies are entering the game as well.

In addition to securing a substantial share of the U.S. wind market, Timken has its eye on a growing customer base overseas. Just a few yards from the defective main shaft bearing, laid out on a platform, is a huge metal ring, larger in diameter than the men standing by it are tall. Tightly spaced cylindrical steel rollers are sandwiched between its inner and outer rims, like tapered teeth in the maw of a large beast. In the coming days, this supersized bearing is to be loaded into the belly of a 747 headed for Beijing. From there, it will make its way by truck to a wind installation in Hunan province, where it will be tested as a prototype on a turbine manufactured by XEMC Windpower. This Chinese company is competing for a share of the market for "direct-drive" turbines—machines that generate electricity directly from the rotor, rather than relying on a main-shaft-and-gearbox configuration to feed a high-speed generator. Large, built-to-last bearings of this sort are at the upper end of Timken's price scale, easily exceeding $100,000 and sometimes reaching $200,000 for a bearing equipped with computerized sensors that provide automatic, temperature-controlled lubrication of all moving parts.

The prototype bearing that we are admiring was produced at a Timken plant in South Carolina, but Jim Charmley and his colleagues have no intention of flying or shipping multiple bearings of this size halfway around the world. If the prototype works to XEMC's satisfaction, production will shift to one of the half-dozen factories that Timken operates in the People's Republic. Already the company's Chinese manufacturing plants supply bearings, lubrication systems, and other related products and services to a number of Chinese heavy industries, and now it is looking to make wind a major part of its

China business. Aside from the Chinese market's unparalleled growth potential, the starkly efficient way China goes about building its industrial and transportation infrastructure makes it an appealing home for Timken manufacturing.

Timken's international marketing team is more than willing to do business in a nation run by central planners, where the state plays a dominant role in governing and, in some cases, owning the companies involved in turbine manufacturing and wind farm development. But here in America, Timken's leaders take a very different attitude toward government intervention in the industrial sector. Ward J. "Tim" Timken Jr., board chair since 2005, is a case in point.

Speaking at Ohio's Ashland University in October 2009, Tim Timken leveled a broadside attack on the Obama administration's economic recovery plan as a threat to American free enterprise. "Not since the Great Depression has government interfered so dramatically and so decisively in the economic life of our nation," he declared, warning that the administration's emergency stimulus expenditures gravely undermined any remaining public confidence in the private sector. "Nothing could be worse for our economy and our nation," he asserted. He has been equally vehement in his assault on legislative proposals that would cap U.S. carbon emissions and require power plants and other major industries to pay for any carbon dioxide (CO_2) emissions above their allotted amounts. He told the Ashland University gathering that this sort of "cap-and-trade" regime would "drive up the price of energy, deter American job creation, and send our jobs overseas."[22]

Timken's leaders have reason to be worried about the financial implications of any regime that attaches a price to carbon emissions. The company's bearings are manufactured from recycled steel, in electric arc furnaces that process the steel-content equivalent of 100,000 cars per month. Timken's Ohio facilities alone consume up to $50 million a year in electricity, mostly generated by coal.[23] Communications manager Lorrie Crum contends that "we couldn't possibly get enough sun or wind here to power our operations without relying heavily on conventional sources."[24]

Crum may be right that, today, her company is more or less stuck depending on coal to fuel its steel plants and other factories. Yet she discounts too readily the longer-term potential for electricity generated by wind and other renewable resources. While Timken's home state of Ohio has installed only a few megawatts of wind capacity to date, plans for building 1,000 megawatts of offshore wind on Lake Erie by 2020 have been announced.[25] To the east, Pennsylvania has 748 megawatts of wind up and running, and New York developers have installed nearly 1,300 megawatts at their wind farms. To the west, wind power in Indiana has already topped 1,000 megawatts, Illinois has more than 1,500 megawatts of installed wind, and major offshore wind projects are now on the drawing boards for Lake Michigan. Reaching farther into the Great Plains, improved high-voltage transmission could open up truly vast resources in the heartland's wind belt, extending from North Dakota and Minnesota down through Iowa and Nebraska to Kansas, Oklahoma, and Texas. With the right policy priorities and market incentives, Timken and other Rust Belt companies could end up substantially curbing, if not eliminating, their reliance on conventional fuels for most of their power needs.

Wherever they lie on the political spectrum, and whatever their attitudes about climate change and the role of government in shifting us away from carbon-based fuels, U.S. manufacturers are finally embracing the new opportunities that wind energy offers to their own business interests and the American economy. The growing pains of smaller companies like Clipper should warn us that we are still in the early stages of developing an entirely new technology base for our nation's energy future. As in any pioneering technology field, the design challenges and operational surprises may stymie us at times, just as the financial setbacks may be harrowing. For decades, the federal government has generously nurtured our fossil fuel and nuclear industries. It behooves us to do the same for wind and other renewable energy technologies that can help create a less precarious platform for future development. With a serious commitment to cleaner energy choices, we can promote a vibrant, sustainable U.S. economy without perpetuating our reliance on foreign energy sources, without exposing ourselves to nuclear energy's hazards, and without mortgaging the global climate.

The challenges facing U.S. wind technology companies are complicated by the fact that they are not alone in looking for ways to consolidate their positioning in the clean energy marketplace. High-caliber equipment coming out of Denmark, Germany, and Spain has long competed for U.S. sales, but with European manufacturers shifting much of their production to America, the net gains to U.S. workers and the U.S. economy are substantial. Much harder to gauge is the role that Chinese wind technology will play in the years ahead. As in so many other sectors, China is a hugely ambitious, rapidly growing force in the global wind trade, benefiting from cheap labor costs and brashly protectionist policies at home as it begins to market its turbines abroad. Although Chinese companies are only making their first forays onto U.S. soil, American wind manufacturers have reason to be nervous.

CHAPTER FOUR

The Chinese Are Coming

LEGEND HAS IT THAT Napoleon once pointed to a map of China and cautioned: "Let China sleep, for when she wakes, she will shake the world." I felt the prophetic impact of those words as I looked around me at the thousands who had gathered for China Wind Power, a conference and trade show in Beijing promoting China's rapidly expanding wind energy industry.

As American companies look for ways to expand their niche in a growing global wind market, China is both an enormously alluring export target and an increasingly formidable competitor. Though a latecomer to the field, the People's Republic is now the world's biggest user of wind energy, each year installing more new wind power than any other nation. Predictably, General Electric, Timken, and other American companies that already have a strong presence in China are revamping their local plants to capture a corner of the Chinese wind market, while other U.S. firms are seeking out Chinese buyers for a dizzying array of wind energy equipment made in the United States. Breaking into the Chinese market is no small feat, however. Protectionist policies give Chinese companies an enormous advantage over foreign competitors—even those that are employing Chinese workers at local factories.

In addition to being the world's biggest consumer of wind energy technology, China is quickly rising to the top ranks of wind equipment manufacturers—largely by virtue of the enormous scale of China's internal market for wind. Four of its companies are now on the "top ten" list of global wind turbine producers, and seven of the top fifteen manufacturers worldwide are Chinese.[1] Perhaps most worrisome to American and European turbine companies, China is selling its machines at cut-rate prices that may soon make them daunting competitors outside its borders. Already a few leading Chinese companies have opened sales offices in the United States, and one turbine manufacturer is building its own commercial-scale wind farm in Illinois, primarily to prove the worthiness of its turbines to American wind developers.

Steve Sawyer, secretary-general of the Global Wind Energy Council and former head of Greenpeace, invoked Paul Revere's memory when he spoke about China's expanding presence in this hotly competitive field. "Ever since China entered the wind market, we've been hoping, anticipating, fearing that 'the Chinese are coming,'" he said at China Wind Power's opening session. Other speakers made more pointed remarks about China's protectionist rules and the special subsidies favoring domestic over foreign manufacturers. Li Junfeng, secretary-general of the Chinese Renewable Energy Industries Association, tried using humor to divert attention from the substance of these claims. "Wind in China is in its puberty," he observed with a laugh. "No one talks about what you do when you're in your fifties or sixties, but at sixteen, everyone's watching."[2]

China's wind energy leaders may function within an industry that continues to be dominated by communist, state-run enterprise, but their entrepreneurial zeal poses a very real challenge to rivals who were raised in the cradles of European and American free enterprise. In wind energy, as in so many other arenas, China is fast becoming a nimble, dynamic and—to many—unnerving force in the global marketplace.

A week before the China Wind Power gathering, I traveled to Tianjin, home to the world's biggest concentration of wind energy manufacturing facilities. Nearly 25,000 people work in the Tianjin wind trade, employed by several of the top global and Chinese wind technology companies. Vestas has the largest presence among turbine producers, initially opening a blade plant there in 2006 and, more recently, adding factories for nacelles, hubs, and generators. I was in Tianjin to attend the opening of the company's two latest facilities at a vast industrial campus that has become the most massive Vestas factory complex anywhere.

An eye-stinging haze enshrouded this city of 12 million as our sleek white bus moved through a dense fabric of elevated truck routes, high-rise apartment blocks, and vast, low-slung factory buildings. This was the China I had long heard about—endless urban sprawl in a fog of filthy air that blocked out the sun and erased all hints of sky. The side panels of our bus, with their photo images of wind turbines against an azure sky, stood in sunny contrast to these bleak surroundings.

At the Vestas manufacturing complex, I joined hundreds of Chinese workers who were filing into a cavernous steel-frame shed to attend the factory-opening ceremony. Dressed in polo shirts with the Vestas logo, they were followed by European company officials in Western business attire who took their places alongside local dignitaries on a makeshift stage. Arrayed behind them was a blue-and-green electronic banner, in English and Chinese, proclaiming the company's mission in China: "Building a World Class Production Base in the World's Fastest-Growing Wind Energy Market."

Lars Andersen, president of Vestas-China, came to the podium. Reading from an English-language text, he rolled out his company's local credentials. Vestas, he reported, was the first company to install grid-connected turbines in China in 1986. Its manufacturing complex in Tianjin has 1.4 million square feet of factory space and cost the equivalent of $363 million to build. Bounding to the mike, the next speaker was the Tianjin industrial zone's vice chairman, Ni Xiangyu, his exuberance palpable as he addressed members of the audience in their native tongue. "This could be the best money-making machine in the world, using the best brains in the world!" he exclaimed.[3]

Ni's brazen boosterism and Tianjin's embrace of wind are local reflections of China's new dominance in the wind energy field. In 2003, the People's Republic brought 100 megawatts of new capacity on-line, little more than 1 percent of the 8 gigawatts installed globally that year. By 2009, it had outstripped all other countries in new installations, capturing 36 percent of the world market. In 2010, it took close to half of the market: its 18.9 gigawatts (18,900 megawatts) of new capacity gave it 49.5 percent of all newly installed wind power worldwide.[4] Vestas had good reason to be locating its largest production facilities in the world in Tianjin.

China still lags far behind the United States and other Western nations in its per-capita wealth, but it is working hard to narrow that gap. At least in its larger cities, the quest for comfort, mobility, and personal expression through fashion has redefined the national ethos. Urban boulevards formerly filled with bicycle commuters are now gridlocked with Volkswagens, Audis, BMWs, Volvos, and Buicks, along with several Korean and domestic car brands. Glitzy shopping malls flaunt luxury designer clothing and accessories. Soaring office and residential towers line the streets, and beneath them run mile after mile of elegant and immaculate subways, with new lines opening every year.

While much of rural China still may be quite poor, the Communist Party and its planners are expanding the economy at a feverish pace. Throughout the first decade of the millennium, China's gross domestic product (GDP) grew at an average of more than 9 percent annually, a rate that is likely to be matched, or nearly so, in the years ahead. In fueling that growth, China has overtaken the United States in its hunger for electric power. By 2020, the country is expected to consume twice the electricity that it uses today.[5]

So what will be wind energy's role in fueling this phenomenal growth? A joint team of researchers from Harvard and Beijing's Tsinghua University has concluded that China's onshore wind resources could supply more than seven times the country's current electricity needs.[6] Yet China is not Denmark, and its central planning body, the National Development and Reform Commission (NDRC), is not prepared to envision a future where wind turbines provide the world's most populous nation with the lion's share of its power.

China's planners are spreading their bets across a broad array of power plants reliant on coal, nuclear, hydroelectricity, and natural gas. Of these, coal has been and will remain the primary fuel. A total of 80 percent of China's electricity today is produced from coal, and with several dozen new plants coming on-line every year, the country's coal-based power production will increase dramatically in the years ahead.[7] Already consuming more than 3 billion tons of coal annually, China accounts for nearly 40 percent of global coal use.[8] By 2015, its demand for coal is expected to top 4 billion tons per year.[9]

The impacts of relying so heavily on the world's dirtiest fuel are legion. Mining disasters in China kill thousands each year, and that toll will inevitably rise as more and more coal is dug from the earth.[10] Much of the air pollution that hangs over whole regions of China comes from coal burning, and that too will increase despite lip service paid to new "clean coal" technology. And burning all that coal will only push China further into the lead as the world's biggest contributor to global warming.

In addition to the new coal plants, China now has 27 nuclear reactors under construction—three times as many as its closest rival, Russia, and nearly half of all nuclear plants being built worldwide. China also dwarfs all other nations in the 50 additional reactors now being planned and the 110 that have been proposed.[11] As if Three Mile Island and Chernobyl didn't give us sufficient warning about the perils of nuclear power in the 1970s and 1980s, we now have the recent disaster at the Fukushima Daiichi reactor complex in Japan to refresh our recollections. It remains unclear whether China will heed this warning. Before peppering the countryside with nuclear reactors, China's central planners would be wise to take more careful stock of the hazards associated with this technology, particularly in light of the catastrophic earthquakes that have recently shaken parts of that nation.

Chinese energy planners are also moving ahead with a commitment to ramp up the country's use of hydropower, ostensibly greener than coal and nuclear in its reliance on a renewable energy resource. However, as the world's largest hydroelectric project moves toward completion at the Three Gorges Dam, along the Yangtze River, the human and ecological effects of giant dams are hard to ignore. More than 1 million people have already been forced from their homes by this project, and by the time it is complete, more than 5 million residents could be displaced.[12]

With so many other energy projects under way or planned, wind energy remains a relatively small part of China's energy portfolio. Maximizing economic growth is clearly the order of the day—far more important to China's political leadership than protecting the country's environment or reining in global warming. Yet China's dramatic progress in increasing its use of wind power reveals what its planners can do when they decide that a given investment is in the state's interest.

~

China's use of wind energy remained minimal throughout the 1980s and 1990s. Despite the government's issuance of a strategic plan calling for 1,000 megawatts of wind power by the year 2000, only 200 megawatts of new wind power had been installed nationwide by 1999[13]—roughly equal to the generating capacity of a single large wind farm in America like Cloud County's Meridian Way.

Beginning in 2002, however, a number of reforms helped bring wind closer to the mainstream. The government that year invited competitive bids for the development of selected wind farm sites, and in just four years, China's installed wind power quintupled.[14] Then, in 2005, the country's first Renewable Energy Law called for mid- to long-term national targets for renewable energy use and for plans at all levels of government to achieve those targets.[15]

This development gave Ren Dongming and his colleagues the mandate they needed to expand wind energy's contribution to China's energy mix. Ren is deputy director of the Center for Renewable Energy Development, a branch of the NDRC that is charged with planning for China's wind energy future. On meeting him at the government-run Guohong Hotel, a few blocks from his Beijing office, I was surprised and relieved. Dressed in blue jeans, a denim shirt, and a tweed jacket, Ren peered at me through silver wire-rimmed glasses, looking to me more like a Berkeley professor than a Chinese government bureaucrat. As we talked over sodas in the hotel's noisy restaurant, his level-headed analysis cast light as well as hope on China's growing commitment to wind.

Ren spoke of China bringing 150 gigawatts of wind power online by the year 2020, or about 8 percent of the nation's projected power-generating capacity. To make this happen, he and his team have mapped out a series of wind farm megabases—vast wind-rich zones where new projects are to be commissioned on a scale unprecedented in China or anywhere else in the world. In eastern Inner Mongolia, one zone has been slated for 30 gigawatts of wind power—substantially more than China's overall wind power capacity at the time that the megabase program was officially unveiled in June 2009. Western Inner Mongolia has been targeted for 20 gigawatts. The Jiquan Wind Farm Base in Gansu province, first of the megaprojects to begin construction in 2009, includes thirty wind farms adding up to 12.7 gigawatts. This project alone, according to official projections, will displace nearly 10 million tons of coal combustion annually.[16] In all, seven megabases have been designated in six Chinese provinces, adding up to 120 gigawatts of new wind-generation capacity at a cost of roughly $140 billion.[17]

With other, smaller wind projects scattered across China and new offshore installations in various stages of planning, Ren Dongming's 150-gigawatt goal seems well within reach. Already the country has dipped its toes in the water with the 102-megawatt Donghai Bridge wind farm just off Shanghai's shores, and the NDRC maintains that offshore wind arrays could provide close to 500 gigawatts of new wind power capacity. That's in addition to the vastly greater wind resources available on land.[18]

Big state enterprises are the predicted winners in the bids for megabase wind projects, partially because they can readily tap the material and financial resources to deliver huge new increments of wind power. At the same time, private investors are finding their way into China's growing wind market. Typically, they are emerging as owners and operators of smaller wind farms that can be approved by local authorities without any national government review. Wind farms with fewer than 50 megawatts of installed capacity fall under local government jurisdiction, free from the NDRC's more cumbersome permit application process.

UPC Renewables is the Chinese subsidiary of a private partnership founded in 1994 by a group of international investors, some of whom were pioneers in the California wind boom of the 1980s. After developing wind and solar projects in Europe, the United States, North Africa, and the Philippines, the firm's managers started looking for a way to break into China's wind energy market. They hired Guo Zheng, an electrical engineer trained at North China Electric Power University with two decades of experience in the oil and hydroelectric sectors.

Like most Chinese nationals who are active in international business, Guo Zheng has adopted a Western name for his dealings with foreigners. He introduces himself as Wilson Guo when I visit UPC's China headquarters, on the seventeenth floor of a downtown Beijing office tower. I ask him to describe how he lines up new projects and am struck by the contrast with Jim Roberts's work as Horizon's land agent in Cloud County. He has no direct dealings with farmers and ranchers; Guo's only business is with local government officials. "Farmers have the right to live on the land," he explains, "but the land is owned by the government."

When the Maoist collective farms were broken up in the 1980s, individual farmers received plots of land that they could either farm for themselves or lease to others. Actual ownership of the land remained with the collectives, however. Today, when a wind developer leases land from a collective, all members of the collective receive a share of the funds. "The government has a standard formula for land compensation so you know what it will cost you from the beginning," Guo says.[19] He compares this favorably to countries like the United States, where wind farms have to be assembled through individually negotiated contracts with multiple landowners.

Others express qualms about Chinese wind farm siting politics. Under-the-table payments to local officials are all too often part of the cost of doing business in rural China, as several close observers told me. Moreover, there is no real opportunity for members of the public to voice their concerns about badly sited projects. Charlie McElwee, a Shanghai-based American attorney and an adviser to wind energy developers, tells me that he hasn't heard of a single wind farm in China being stopped or altered because of concerns about migratory birds, human health, noise, or aesthetics. "Tame" is the word McElwee uses to describe China's environmental movement, where truly independent, citizen-based organizations simply don't exist. He can recall a few recent local demonstrations protesting major industrial projects—a waste incinerator in one locality, proposed petrochemical plants in a few others—but government security forces broke them up quickly. Beyond the local level, he says that the environmental movement generally consists of groups that are under the close supervision of Chinese governing authorities. He calls these groups "GONGOs," or government-organized non-governmental organizations, to distinguish them from the NGOs, or non-governmental organizations, that operate with a high degree of independence in less-authoritarian societies.[20]

Foreign investors like UPC are by no means alone in acquiring a stake in China's local and provincial wind farm development. Tianrun New Energy Investment Company, Ltd., is the Beijing-based development arm of Xinjiang Goldwind Science & Technology. Goldwind was an early pioneer in China's wind industry, producing its first turbines in 1998 under license from a German manufacturer. This practice of piggybacking on foreign turbine designs became the industry norm among Chinese manufacturers, allowing them to enter the market quickly, though not necessarily with the most updated technology.[21] By 2007, Goldwind had become the country's market leader in turbine production, and it began investing in wind farms through Tianrun. Within two years, it had gained full ownership of eight wind farms and held a controlling stake in six others.[22]

Unlike UPC, which is a privately held foreign enterprise, Goldwind is a hybrid typical of companies operating on the seam between China's old-line communist state-owned enterprises and the new generation of private companies that are finding their place in the country's emerging market economy. The company, on the one hand, is majority-owned by private investors; its shares have been publicly traded on the Shenzhen Stock Exchange since 2007. On the other hand, several of Goldwind's biggest investors are state-owned enterprises; Tianrun officials estimate that 30 to 40 percent of the company's assets remain in state hands.[23]

With multi-gigawatt megabases already under development and many companies like UPC and Tianrun moving forward with smaller projects, China's new wind capacity is growing at a breathtaking speed, accelerated by the ease of lining up land parcels on communal land, the very limited focus on environmental concerns, and the absence of any real public input into the siting and permitting processes. The contrast with America's methodical and highly participatory approach to project development couldn't be starker. Not surprisingly, transmission companies are having a hard time keeping pace with so much new, dispersed power production. In recent years, as much as a third of the country's installed wind power has not made its way onto the grid and either stands idle or remains landlocked in local areas that often can't make good use of the available power.[24] The government has committed to build a vast new network of transmission lines, but with overall Chinese electricity use expected to nearly double by 2020,[25] grid operators are in an ongoing race to keep up.

Ensuring a steady supply of reasonably priced turbines to new Chinese wind farms is another challenge the government has addressed, often in ways that have aroused the ire of foreign turbine manufacturers. In the early years of China's wind power development, foreign companies supplied the lion's share of the technology. Today China's domestic brands dominate the market.

Highly protectionist regulations have given Chinese manufacturers a decisive edge. Starting in 2003, local manufacture was made a key criterion in the review of bids for government-sponsored wind farm concessions. At first the minimum "local content requirement" was set at 50 percent of a turbine's value, but a year later it was raised to 70 percent. Soon this protectionist regime extended beyond government projects to include all turbines installed in the country.[26]

Foreign manufacturers that relied heavily on imported components had a tough time meeting the local content hurdle. In 2004, international brands commanded three-quarters of turbine sales. Within five years the proportions had more than reversed, and by 2009, Chinese manufacturers had captured 86 percent of new installations.[27] By 2010, the only non-Chinese manufacturers to make it into the top-ten ranking were Vestas, with less than 5 percent of the market, and Spain's Gamesa, which accounted for 3 percent of new wind capacity. General Electric, the sole U.S. player of any size in China, had a market share that barely exceeded 1 percent.[28]

For years the United States had complained about Chinese anti–free trade measures, but in 2009 the Obama administration singled out the local content requirements on wind turbines. While U.S. companies like Timken could sell bearings made at their Chinese factories to local turbine producers, dozens of other U.S. companies, producing their wares on U.S. soil using American workers, were effectively blocked from participating in the Chinese wind energy boom. China, under pressure, signaled its readiness to lift the local content requirements, but a year later the U.S. Trade Representative petitioned the World Trade Organization (WTO) for a special consultation on a broader range of policies that were seen as giving Chinese wind turbine producers an unfair edge over foreign competitors.[29] Not only were the Chinese still making it hard for U.S. goods to enter the Chinese market, but leading Chinese turbine manufacturers were preparing to export their own technology to America. This, no doubt, raised the anxiety levels of U.S. officials, as well as the United Steelworkers, who instigated the WTO petition.[30]

Chunhua Li, director of international business at Goldwind, made it clear to me that America is a prime target for Goldwind sales, when we met at the company's research and development center in an industrial park just south of Beijing. His words reflected cautious determination: "We want to be sure that our turbines supplied to overseas developers make our international market successful. Now we feel we are ready." He proudly described the "green-friendliness" of Goldwind's direct-drive turbines—simpler to build and easier to maintain than conventional gearbox-dependent machines. He also pointed to Goldwind's newly introduced 2.5- and 3-megawatt models as proof that the company can match the scale of equipment now being sold to the U.S. and European markets.[31]

After our meeting, I was taken past the company's soccer field and climbing wall to one of Goldwind's turbine assembly plants. Li Fan, a youthful-looking production manager, greeted me at the entrance. Eager to practice his English, he took me on a tour of the production room where 1.5- and 2.5-megawatt turbines were being assembled. Giant rotors and stators, the revolving and stationary parts of direct-drive turbine generators, were laid out on heavy wooden blocks in one quadrant of the factory. In another area, workers were carefully installing electronic controls inside turbine nacelles, and in a far corner of the factory, assembled generators were being tested on a full-scale turbine simulator.

Li Fan had good reason to brush up on his English. In only a few days, he would fly to America to oversee the erection of three Goldwind turbines at a demonstration site near Pipestone, Minnesota. By selecting a site exposed to some of the harshest weather conditions in the lower forty-eight states, Chunhua Li and his international marketing team hope to persuade American wind developers that Goldwind's turbines can outperform better-known Western brands. As a further step toward making its technology known, Goldwind is now building a full-scale commercial wind farm in Lee County, Illinois. The company's U.S. affiliate, Goldwind USA, has already secured a twenty-year commitment from Commonwealth Edison to buy all the power generated by this 109-megawatt farm.[32]

Goldwind's strategists are competing fiercely with other Chinese companies that now have set their sights on the global market. The most formidable among them is Sinovel. Barely noticed in 2004, when it was launched as a subsidiary of the state-owned steel giant Dalian Heavy Industry Group, Sinovel entered the market two years later with a 1.5-megawatt turbine produced under license from a small German company, Fuhrländer. By 2008, it had eclipsed Goldwind to become the leading turbine supplier to the Chinese market, and by 2010, it held 23 percent of total domestic installations, compared to Goldwind's 20 percent share. Next in line were three other Chinese manufacturers; then came Vestas in sixth position, holding less than 5 percent of the Chinese market.[33]

Sinovel's publicists are brash and uncompromising in their reach for global supremacy. Through a "Three-Three-Five-One" strategy announced in 2009, the company unveiled its ambition: within three years, to become one of the world's top three turbine manufacturers; and within five years, to be number one in the world.[34] The company exceeded its first milestone in only a year, edging out General Electric to become the number-two global manufacturer in 2010. With 11 percent of all new installations worldwide, Sinovel ran just a single percentage point behind Vestas. General Electric and Goldwind came in third and fourth, each with about 10 percent of the market.

Exports remain a minor part of China's turbine manufacturing success story, yet the country's domestic market is big enough to create a hugely robust wind industry. In 2010, with nearly half of the world's new wind power installations taking place on the home front, China's turbine producers have been able to project a dominant presence on the global stage, even though their physical reach has largely remained within a single country's borders.

Chinese entrepreneurs are bullish about exploring new frontiers abroad, but it remains unclear whether their turbines will cut significantly into the European and U.S. markets where familiar, well-proven brands now have the edge. Industry analyst Matthew Kaplan describes Chinese turbines as based on borrowed "tier three" designs that are typically several years out of date, causing them to be less productive than the models that "tier one" Western manufacturers are placing on the market.[35] In determining just how much of a bargain they are getting from a price discount on Chinese turbines, wind developers will need to weigh very carefully the full life-cycle performance of the equipment they are purchasing. Developers building wind farms in America will also have to reckon with the higher transportation costs of importing equipment from halfway around the world, and with the difficulties they may face raising capital for new wind farms using lower-quality turbines. Finally, the price advantage of Chinese hardware is being weakened by a drop in the price of American-manufactured turbines, caused by surplus capacity among existing suppliers to the U.S. market.

Regardless of how effective Chinese companies turn out to be in breaking into new wind energy markets, China's planners and policymakers have cleared a startlingly expeditious path to developing their own country's wind energy potential. With more wind power installed in China than in any other nation today, the National Development and Reform Commission projects that the country's wind installations could reach 1,000 gigawatts by 2050. That's five times current global wind-generating capacity (see table 5) and would be enough, the NDRC estimates, to provide about 17 percent of China's total power needs that year.[36] If the recent past is a reliable predictor, the Asian giant could end up far exceeding this goal well before 2050. It certainly has the wind resources to do so.

China's ability to match concrete programs to bold ambitions is practically unparalleled, but these transformative powers come at a high price. Private property rights scarcely exist. Government policies and plans are announced, not debated. And very few dare to protest government decisions on the streets or in the courtroom. *New York Times* columnist Thomas Friedman has acknowledged his own ambivalence about China's staggering ability to get things done in *Hot, Flat, and Crowded,* his manifesto for a newly defined green revolution.

Decrying the political gridlock that prevents America from making essential, fundamental shifts in its infrastructure and economy, he writes: "If only America could be China for a day—just one day. *Just one day!*" During that day, he dreams of our government being able to cut through the "legacy industries, . . . the pleading special interests, . . . the bureaucratic obstacles, [and] . . . the worries of a voter backlash" that prevent us from reshaping the way we use our resources and treat our environment. But then, once that day is done, he wants a vibrant and vigilant public interest sector to step back in to make sure government agencies and private businesses fulfill their responsibilities under the new order.[37]

Friedman's hybrid of strategically focused authoritarianism and participatory democracy may be a pipe dream, but his invocation of China is intended as both a lament and a challenge to Americans. He wants us to reckon honestly with just how hard it will be for our nation to make transformative decisions that will truly serve the long-term welfare of our nation and planet. At the same time, he is daring us to summon the political and civic courage to bring about those very changes.

On the long flight back from Beijing, I prepared myself for the next stage of my research. I would be going into the field to meet some of the thousands of people now working to build America's wind farms and keep them running for decades to come.

CHAPTER FIVE

Working the Wind

BILL STOVALL LOOKS LIKE the kind of guy you'd expect to see blazing through town on a Harley chopper. His head wrap bears a grisly skull-and-crossbones design, and a tangled thicket of white hair bursts from its fringes. Opaque sunshades, lambchop sideburns, and a walrus mustache frame a full, reddened face. His forearms are a splotchy brown mosaic, pummeled by years of exposure to the sun.

In fact, Bill's motorcycle sits in the garage at his home in Gatesville, just west of Waco, Texas. "I ain't rode it in a year and a half!" he tells me with a low, rumbling laugh. Bill is a long-haul trucker, one of hundreds now delivering turbine components to wind farms across America. He works for Lone Star Transportation, a company based out of Fort Worth whose trucks feature telescoping trailers for transporting blades, low-riders that allow giant tower segments to slip under highway overpasses, and multi-axle decks that can carry nacelles, each weighing a hundred thousand pounds or more.

Bill and I meet at a railroad yard in Reynolds, a small town in northwest Indiana. This is the depot where blades, tower segments, and other turbine components are being held until they're needed at the construction site for the Meadow Lake Wind Farm, a 200-megawatt power plant that may eventually be expanded to five times that size. Bill's trailer is loaded with a Vestas V82 blade, 130 feet long, that stretches his truck-trailer combo to 168 feet—well over twice as long as a conventional semi-trailer. He pulls to the edge of the access road, steps down from the cab of his bright-red Peterbilt truck, and walks to the spot where a heavy steel rod, the "kingpin," holds the trailer in place. Turning a crank, he lowers two steel legs so that he can yank the kingpin and "dolly down" the trailer, leaving it standing overnight while he drives off to get a decent night's sleep at a nearby motel. Tomorrow morning before seven o'clock, he will come back to pick up the blade and deliver it to the Meadow Lake Wind Farm, twenty miles away in a flat stretch of farm country just north of Lafayette.

Now in his late fifties, Bill has been a long-haul trucker for years. He used to deliver fresh meats and produce, but for the past few years he has worked in the wind industry. The hours are long, and the time away from home is endless. It's late July, and he tells me it's been sixteen weeks since he's had a single day at home with his wife. Before that, he was on the road for nine weeks. "So I've actually been gone twenty-five weeks with one day at home," he says. And last year was worse. "I drove from May to December and never got home." His wife visited him twice on the road. "That didn't hurt," he says with a smile that vanishes as quickly as it appears. His kids are grown, but he has five grandkids and one great-grandchild; rarely does he see them.

What keeps Bill going, along with the wages, is pride. When he was delivering meats and produce, he saw himself as performing an important, if little-recognized, service. "You go into the stores now and everyone wants fresh lettuce and eggs and fresh meat, and there's a whole lotta old boys out on the road that go to a lot of trouble on a regular basis to make sure that stuff is there every day." There's dignity, not bitterness, in his tone.

Wind energy brings Bill equal pride. "When I go by these wind farms, I think, 'I did my part of that. I can *see* that.'" He's boned up on the basics of how turbines work, the power they produce, the fossil fuels they displace. "I found out [one turbine] puts out electricity for a thousand homes, pays for itself in five years, and lasts fifty—not a bad deal!" His reasoning is plain: "We don't buy the air. Somebody might start chargin' for it, but for now, it sounds better 'n coal."

The next morning, I set out from my motel at 6:30. My goal is to catch the nine-vehicle convoy that Bill will be joining: three trucks carrying one blade each, plus six escort vehicles—one in front of each blade truck, one behind. Blades travel in matched sets from factory to wind farm, their weights carefully measured to make sure that every turbine has a well-balanced rotor.

The fog is so dense I can barely see fifty feet ahead of me as I make my way south on I-65. Eventually I spot blinking orange roof lights on a green Ford pickup truck in the right-hand lane. Ahead of it, I make out the contours of a giant white blade. Moving into the left-hand lane, I edge by one trio of trucks—escort, blade, escort. Then I pass a second trio, then a third. The caravan is cruising at a surprisingly swift clip given the poor visibility: about fifty miles per hour, just slightly slower than the general traffic.

Bill calls the forward escort for his blade truck his "front door" and the rear one his "back door." The back door has the tougher job: its driver has to steer the rig's rear axle remotely from a few dozen feet behind the truck, using a handheld electronic device that looks like a simplified TV control, all the while keeping his own vehicle on the road. It makes texting while driving seem easy.

The advantage of having a double-escort becomes obvious when we pull off I-65 and negotiate a tight left turn onto an overpass heading east on State Road 18. Before the blade truck enters the turn, the front-door driver hops out of his minivan and uproots the stop sign at the end of the exit ramp. Otherwise the blade's tip would level it on its wide swing around the ninety-degree turn. The back-door driver then navigates the turn, keeping the blade truck's rear axle heading straight while the cab veers off to the left. Once the last blade has made its way around the turn, the trailing escort replants the stop sign.

A few miles down the road, the three blade trucks followed by their rear escorts peel off the pavement to the right, edging slowly onto a semicircular dirt path that allows them to make another ninety-degree left-hand turn. Completing the arc, the trucks head off into a fogbound sea of shoulder-high green corn. The front escorts, unneeded at this stage, stand idle in a grassy holding area while the trucks unload their wares at the foot of a nearby turbine tower.

A friend who has dug into National Archives photos from the Cold War tells me these wide-swinging turnoffs remind him of the aerial surveillance shots that tipped off U.S. intelligence to Cuba's installation of missiles in the early 1960s.[1] There were no giant wind turbines then, so what else would Fidel Castro have been moving around the countryside?

For the blades and other turbine components that have made their way to the Meadow Lake Wind Farm, the 25-mile ride from the Reynolds railroad depot is the final short leg of a very long journey. Some of the blades have come by truck from a Vestas factory in Windsor, Colorado, 1,000 miles due west. Others were shipped out of Aarhus, Denmark, destined for the Port of Burns Harbor, Indiana, on Lake Michigan's southern shore. The journey across the Atlantic, down the St. Lawrence Seaway, and through four Great Lakes covers 4,500 nautical miles and takes about fourteen days. Although each load is different, the ship that brought in Bill Stovall's cargo carried 94 blades, 30 power generators, and 30 hubs.[2]

Arduous though shipping from Europe may be, the hands-down long-distance medalists at Meadow Lake are the towers. Purchased from a supplier in Vietnam, they were shipped in sections across the Pacific to the Port of Vancouver, Washington, on the Columbia River just north of Portland, Oregon. That leg of the trip racked up well over 6,000 nautical miles. The next leg took them on flatbed railcars cross country to Indiana—about 2,200 miles. At the Reynolds rail yard, I watched as twinned pairs of tall cranes gingerly lifted these multi-ton tubes off their flatbeds and aligned them in neat rows down the full length of the yard. Hundreds of tower sections rested on heavy-gauge steel cradles, each stamped in bold black lettering: "Made in Vietnam."

With increasing numbers of companies—foreign and domestic—opening up manufacturing and assembly plants on American soil, trips carrying turbines from factory to farm may soon be shorter. That comes as good news to equipment manufacturers and wind farm developers alike, as transportation accounts for about 10 percent of the cost of building a U.S. wind farm.[3] But for now, wind energy is big business for shipping companies and the ports they use. Ports up and down the West Coast, from Longview, Washington, to San Diego, handle massive shipments coming from Asia.[4] Along the Gulf Coast, five Texas ports—Beaumont, Corpus Christi, Freeport, Galveston, and Houston—serve as wind technology gateways.[5] And in the Upper Midwest, the Great Lakes ports of Burns Harbor and Duluth enjoy an increasingly brisk wind trade. Ports along the eastern seaboard are not yet major players in the wind industry, but that may change if offshore wind farms are built in the Atlantic, as is expected in the coming years.

Railways are also busy moving turbine components from ports and factories to their ultimate points of use. The Union Pacific is already moving thousands of carloads of wind turbine components from ports and assembly plants to wind farms annually, and it is prepared to handle up to 6,000 railcars loaded with wind equipment at its logistics center in Manly, Iowa.[6]

Though rail haulage is gaining momentum, trucks will remain at the center of U.S. wind technology transport. Needed, in any case, for carrying turbine components their final miles to construction sites amidst cornfields and cattle pastures, they are less cumbersome than rail for most trips shorter than several hundred miles. Shifting heavy loads from one transportation mode to another simply isn't worth the cost, time, and effort if the distances aren't truly great. Even when land travel is long, wind developers often opt for trucks, especially if they are carrying big-diameter tower segments that can't make it through railroad tunnels or super-long blades that require the use of two specially fitted railroad cars per unit. Looking at land traffic coming out of the Port of Longview, the proportions are telling: 90 percent of the turbine components leave by truck.[7]

All of this gives plenty of work to people like Bill Stovall, who are now busy, year in, year out, moving wind energy equipment from ports, railroad depots, and factories to wind farms across the country. Beyond the haulers, thousands more have found work at wind farm construction sites, readying access roads, pouring turbine foundations, erecting towers, and installing electrical systems. And then there are the thousands of people who will operate and maintain the wind farms for decades to come.

—

I arrive at Meadow Lake, Bill Stovall's destination, to find half the wind farm's 121 turbines already standing tall. Meadow Lake's developer, Horizon Wind Energy (the same company that built Meridian Way) has dispatched two familiar veterans to supervise construction: Carole Engelder and Alvin Cargill.

Carole flew in last night, held meetings all morning, and slipped away at lunchtime without my catching so much as a glimpse of her. We had planned to meet here, but I think she has come to regard me as one of the "gnats" she complained about back in Kansas, getting in the way of expeditious project development.

Alvin, Horizon's on-site manager, is more courteous. He takes a few minutes to speak with me in the company's field office, a minimally furnished trailer that sits in a row of identical white mobile offices housing the administrative staffs of the half-dozen firms most directly involved in building the wind farm. The trailers are lined up like dominoes at one end of the laydown yard, a graded dirt expanse where concrete is mixed, gravel and other construction materials are stockpiled, and dozens of construction crew pickups are parked alongside fuel trucks and road graders. Hanging from a sturdy chain-link fence surrounding the yard is a banner displaying the general contractor's slogan: "WORK SAFE! Your family needs you." These words remind me of a visit I paid to another construction site, at the Grand Ridge Wind Farm just outside Marseilles (pronounced mar-SALES), in Illinois. There I stood in spring rain alongside Adam Hartman, the fresh-scrubbed site manager for Invenergy, a Chicago-based wind developer. Almost directly above our heads, a 42-ton tower segment was dangling from a 300-foot crane. Adam assured me that serious accidents are rare at wind farm construction sites, but he indulged in a bit of gallows humor as he touched the plastic rim of his hard hat. "We call these brain buckets," he said, explaining with a smile that they're particularly useful in scooping up cranial matter in the event of a major mishap.[8]

About 350 workers are involved in building the Meadow Lake Wind Farm. Alvin breaks that number down for me. Bowen, the general contractor, has just over a hundred people on site doing "civil" work. Along with grading, widening, and reinforcing the access roads that are spread across tens of thousands of acres, they have poured all the foundations for the turbines.[9] An electrical subcontractor, Hinkels and McCoy, employs about the same number, digging tens of miles of trenches and laying electric cables that will eventually carry power from turbines to the grid. Once the civil and electrical infrastructure is in place, Barnhart Crane & Rigging comes in with about ninety workers to do the heavy lifting—stacking and securing the tubular segments of each steel tower, topping off each tower with a nacelle, and hoisting three-bladed rotors into position. Vestas is also on site with a few dozen technicians who make sure the turbines' mechanical, electrical, and hydraulic systems are working properly.

Pouring foundations not only demands a sizeable workforce; it also consumes formidable quantities of concrete and steel. At Meridian Way, I recall being told that each foundation contains 525 cubic yards of concrete and 40 tons of steel rebars, woven and then poured in the shape of a giant inverted mushroom. Emerging from each concrete base are the threaded tips of 144 eleven-foot-long bolts, arrayed in a dense double-ring that will be used to secure the lower tower section.[10] Between these enormous bolts and the hundreds of shorter ones that hold upper sections of the turbine together, the American Wind Energy Association (AWEA) estimates that in 2008 alone, wind developers used 2.4 million oversized bolts to erect about 5,000 turbines.[11] No wonder Cardinal Fastener, Ohio's custom-bolt manufacturer and home of the "Obama Bolt,"[12] sees wind energy as such a promising new market!

Alvin arranges for me to spend the morning with Steve Maples, site manager for Barnhart Crane & Rigging. In the Barnhart trailer, Steve introduces me briefly to Clint Newbold, the company's on-site quality assurance manager. A former Marine, Clint grew up on a ranch in South Dakota and was working for a county road department when Barnhart hired him. He tells me the company recruited heavily from his corner of South Dakota. When I asked why, he belts out his answer: "Because we're good! We're hard workers. We like to get the job done. We're no-bullshit individuals."[13] With eight people assigned to his staff, Clint keeps close tabs on every Barnhart work crew operating at Meadow Lake, making sure they adhere to a rigorous set of performance guidelines.

Steve and I then hop into his pickup and set out through croplands on a matrix of ramrod-straight, unpaved county roads. Filling our field of vision are dozens of turbines, all planted in unwavering lines running along an east-west axis. Some are fully erected; others are white steel stumps waiting for upper tower sections, nacelles, and rotors to be lifted into place. Beginning our drive at row D, we head south at a good clip, leaving a wall of light-brown dust in our wake. Fifteen minutes later we arrive at row K; this first phase of the wind farm continues through row Q. I can only imagine how vast the fully built wind farm will be.

Phase I of Meadow Lake will deliver 200 megawatts of wind power to the grid—enough to meet the power needs of about 55,000 Indiana households. Horizon's land man, Martin Culik, tells me that the project could ultimately reach 1,000 megawatts, supplying 275,000 homes with clean energy.[14] At that scale, it would be the largest wind farm in the state by a wide margin, stretching across 130,000 acres, covering a quarter of White County and spilling over into Benton County, to the west.[15] Flat as far as the eye can see, virtually all of this land today is planted in corn and soy. In the near future, it will be corn, soy, and windmills.

As he turns off the county road onto a dirt path through the fields, Steve tells me that Barnhart lured him out of retirement back in 2005. He will only volunteer that he's "well past fifty," but my guess is he passed that milestone many years ago. Barnhart first hired Steve to oversee construction at a wind farm in upstate New York; he had previously managed projects for a mechanical and industrial contractor in Tennessee. He's happy to be working again, and he's pretty sure his wife is happy to have him out of the house. "It gives her a break from me," he jokes, but then his tone turns earnest: "You spend all your life workin' and all of a sudden you're not, and all your friends are still workin'. You need somethin' to do."[16]

Supervising Barnhart construction crews more than answers that need. Cranes are costly to rent, and keeping workers fed and housed runs up a hefty tab as well. All of that makes six-day workweeks routine, sometimes including double shifts and occasionally even spilling over onto Sundays. "Some days I wonder what I'm doin'," Steve admits, but he quickly adds: "Other days I know, after a few weeks, I'd be extremely bored [sitting at home]. . . . All those years dealin' with plant managers and workers. I need that camaraderie in my life."

The close company Steve enjoys is palpable as we sit in his truck and watch one of his Barnhart crews "fly a rotor," or lift it into place, connecting it to the turbine's main shaft atop 255 feet of steel tubing. It has just started raining. Sheets of water are pouring off the corrugated metal roof of a nearby barn, but the wind is calm and there's no lightning in the area, so the work goes on.

A dozen men team up to fly the rotor, which has been preassembled on the ground by bolting three blades to a hub. (Barnhart has no women on its field crew at Meadow Lake, though Steve explains that's only because the Ironworkers Local hasn't sent any along.) Four men are on the ground, and four more are atop the tower. It takes a crew of two—a Barnhart operator and a union apprentice called an "oiler"—to manage the Manitowoc "triple 9" crane, mounted on tank tracks with a 300-foot, elbow-jointed boom. A single operator runs a smaller wheel-mounted crane, used to stabilize the rotor as it's lifted. And then there's the quality assurance supervisor from Clint Newbold's shop. "It's A-A-A-L-L-L about teamwork," Steve yells to me above the crackling of his two-way radio and the revving of crane engines. "Nobody's gonna like everybody, but you work as a team. You're only as good as the people you're surrounded by."

This seems particularly true when you're raising a 46-ton rotor that's broader in diameter than a Boeing 747 is long. A lot can go wrong if any member of the team isn't properly trained or is less than 100 percent alert. Most vulnerable, perhaps, are the crewmembers high in the tower who must guide the rotor, foot by foot and then inch by inch, as it approaches its docking position at the outer tip of the main shaft, protruding just slightly from the front of the nacelle.

The Manitowoc's engine howls as its cable becomes taut, tugging on the upper rim of the rotor's hub. Two blades rise, forming a perfect V high in the air. The third blade points straight downward, the smaller crane holding its tip just a few feet off the ground. A few dozen yards away, two men stand in a cleared stretch of cornfield, each grasping a rope looped around the tip of an elevated blade. With these slender tools, they hold the rotor steady as it rises into the air slowly, majestically. This is the largest kite I've ever seen flown.

Twelve long minutes into this operation, the rotor hovers with its hub slightly higher than the nacelle. The rain has let up, and Rusty Fitz gets ready to guide the rotor through its final stages of flight. He is visible high above us, ant-sized, peering over the front edge of the nacelle like a sailor looking out from a ship's prow. He begins issuing radio instructions in a Louisiana drawl: "Luffin' up, brother. Luffin' up." Rusty wants the crane to lift its upper boom slightly, creating a little more space between rotor and mount. Thirty slow seconds pass as the crane operator gently shifts the rotor's position.

Rusty then calls out to one of the ropers in the cornfield clearing. "Comin' in on my left tag," he says, looking for a bit less slack in the line. Another twenty seconds pass. "Alright on my left tag, comin' in on my right." The other roper complies. Then to the crane operator: "Gimme a little cable down. I want 'bout eight, nine feet of cable down." The crane lets out some cable and the rotor descends. After a "skootch" to the right and another to the left to finalize the alignment, the first bolts are fastened between the main shaft and the rotor.

Stepping out onto the hub, Rusty disconnects the crane's cable. Even with a safety harness, this is no job for the acrophobe. As the freed cable rises, he sounds euphoric: "Alright there, Big Daddy, all clear. FINE JOB! FINE JOB!"

With the rotor flown, Steve takes me back to the laydown yard in his pickup. Next in line at the turbine site, Hinkels and McCoy will come in and hook up all the electrical systems. Then the Barnhart quality assurance team will make sure everything's ready for a walk-through by the owners. In a few months, he expects all 121 turbines to begin sending wind-generated power to the grid.

Steve doesn't sound overly sentimental when he talks about his "five aggravatin' grandchildren." He admits, though, that he's proud to be playing a role in opening America's horizons to wind, and he hopes that someday his grandkids will say, "Grandpa had somethin' to do with it."

Erecting turbines has its challenges and risks; so does maintaining them. Those looking for a cushy desk job need not apply for any of the thousands of operations and maintenance—"O&M"—jobs that are opening up at America's wind farms. I learned this the hard way during an earlier visit to Cloud County.

To experience what wind technicians undertake a few times daily, I felt it was important that I try climbing a turbine—straight up a series of narrow aluminum ladders inside the tower to the hatch on the underside of a Vestas V90 nacelle, 262 feet in the air. My folly should have been obvious from the start: I am mildly claustrophobic, vaguely uncomfortable with heights, not particularly robust, and, to echo Steve Maples, "well past fifty."

Before setting out on the climb, I was given basic instruction by Meridian Way's operations manager, a two-tour Iraqi war veteran named Justin Van Beusekom, burly, sun-baked, his lower lip bulging with chewing tobacco. First, Justin had me review a multipage release form outlining the hazards of working up-tower, from hypothermia in the winter to dehydration and heat exhaustion in the summer. Then he fitted me with a hard hat, protective glasses, gloves for a better grip, and a harness heavily laden with clips and straps. I was already wearing the mandatory hard-toed work boots, purchased the previous day at Walmart.

As we drove to the turbine, Justin distracted me by talking about two of the local wildlife dangers listed on the release form: rattlesnakes and the small but surprisingly venomous brown recluse spider. With a slight flair for exaggeration, he warned that the spider's bite can cause a finger to "swell up and explode." As for rattlers, I needed no convincing: the previous day, I had seen one slither across a packed-dirt farm road just a few feet in front of my car. In truth, though, my mind was on neither spiders nor snakes; I was girding my nerves for the climb.

At the site, Justin keyed in a command to pause the turbine's operation; no one goes up-tower when the rotor is turning. We entered the tower through a rounded, shiplike steel door, a few steps off the ground. Inside the air was dank with the smell of grease and plastics. Following Justin's instruction, I latched my harness, with its fall-arresting clamp, to a cable running up the center of a short stretch of ladder to a grated steel platform just overhead. Getting to that first platform was easy. Then came the next span—about fifty feet straight up into near-total darkness. And beyond that, I knew, were two more spans covering the remaining distance to the top—nearly 200 feet into a cavity that would be shrinking from 12 feet across to only 7.5 feet in diameter at the top. I also knew that magnets were all that held these slender vertical spans to the tower's interior wall—a means of preserving the maximal strength of this giant steel tube.

With every step I took, the temperature seemed to rise and the air became heavier, more stagnant. For several hours the sun had beaten down on the tower's exposed steel, warming it through and through. I tried not to look up; I tried not to look down. My arms strained, my thighs ached, my heart was pounding. Panic overcame pride; I knew I couldn't make it.

I half-expected a soldier's scolding when I called down to Justin to say I'd had enough. Instead, he calmly coached me through the mechanics of the descent. As I exited the tower through its narrow hatch, I felt a huge rush of relief that left little room for humiliation.

Back in Horizon's field office, I untangled myself from the web of harness gear and shared with Justin what my two teenaged daughters had said when I told them about my planned climb. "Are you *crazy*?" one exclaimed. The other insisted: "You've *got* to do it if you're writing a book about wind." Justin smiled and responded: "They were both right!"[17]

My awe for the rigors of O&M work only deepened on my visit to the Grand Ridge Wind Farm outside Marseilles, Illinois. Leo Jessen is wind developer Invenergy's point man at this 99-megawatt project, supervising a crew of six O&M technicians. A clean-shaven head and fashionable earring somewhat mask his years, but at forty-seven, Leo is older than the other members of his team. Though still relatively quick at climbing turbines, he pithily observes, "There's old climbers and bold climbers, but no old bold climbers." It's all about pacing, he claims. "You might be in the greatest shape in the world; you might be able to run up the ladder; but you've got to understand that, when you get to the top, you've got to do the work." A young buck who races up the ladder in five minutes and then collapses in exhaustion for forty-five is less valuable, he says, than someone who takes twenty minutes to make the climb but then gets right to work.[18]

Ladders on the GE turbines at Grand Ridge are equipped with counterweights that can reduce technicians' effective climbing weight by as much as 50 percent. Leo stresses, though, that technicians typically carry at least forty extra pounds in tools and safety gear, so getting to the top of a tower still demands real effort.

Weather extremes, up-tower as well as en route, impose further physical stress on O&M technicians. Mechanical heat quickly dissipates when turbines are shut down, leaving workers aloft in winter temperatures that can easily drop below zero. In summertime, the temperature inside a nacelle—ventilated but not air-conditioned—often tops a hundred degrees. "If you're a TV dinner, this is the ultimate job for you," Leo says. "You'll freeze in the winter and cook in the summer."

What do O&M technicians cope with, aside from the weather? First of all, turbines—like any machine—require scheduled maintenance. Gearboxes need periodic oil changes, using specially equipped pump trucks to siphon off the old oil and pump in the new. Grease must be applied to the gears that control the pitch of turbine blades and the mechanism that changes the "yaw," or compass orientation, of the rotor, both of which are constantly adjusting to optimize rotor speed under different wind conditions. Also needing periodic checking and replacement are brake pads, used along with blade-feathering to hold the turbine still during maintenance and, in a pinch, to help keep the rotor from spinning out of control. Then there are the gearbox failures, generator misalignments, electrical defects, hydraulic control issues, and other non-routine problems that have to be handled as quickly as possible to minimize the time and cost of taking turbines out of production.

Leo's goal is to keep the Grand Ridge turbines operating as close to 100 percent of the time as is physically possible. In its first half-year of operations, the turbines performed at 98 to 99 percent availability—the term used to reflect the percentage of time that a turbine is fully operational, or *available* to capture the wind.[19] Leo attributes this very high availability to the fact that the equipment is still new, but it also reflects the responsiveness and training of his O&M crewmembers.

Demand for well-trained O&M technicians, on-site managers, and administrative support staff at American wind farms has been growing, but the rate of that growth has been directly affected by fluctuations in the U.S. economy. In 2009, new wind installations totaled nearly 10,000 megawatts. Although the economy had already slumped, turbine orders placed before the recession allowed the industry to maintain a brisk rate of development that year. To service the turbines that went on-line in that year alone, AWEA estimates that 1,000 new O&M jobs were created.[20] The recession took its toll on wind development the following year, however, cutting newly installed wind capacity in half and substantially reducing demand for new O&M hires. Wind turbine technology has also become more reliable with every passing year, making it possible to spread O&M crews more sparingly across ever-larger turbine arrays.

Even with these market fluctuations and technology developments, the need for trained technicians has been big enough to nurture a new generation of specialized training programs at more than two dozen centers of learning across the country.[21] Cloud County Community College's story is emblematic.

When Bruce Graham gave up his high school teaching career to launch Cloud County's wind program in 2007, he had to work hard at building faculty support. With only four students in his program, he had an uphill burden of persuasion. Before long, though, the school's top administrators came to embrace Bruce's effort as their poster child. Students were coming from several states across the lower Midwest, and money was flowing in—about $1.8 million in federal grants and nearly $1 million in state funding.[22] "Wind has become our icon," the college's vice president for academic affairs proclaimed in March 2010.[23] Today, Cloud County's Wind Energy Technology Department has more than a hundred students enrolled in its two-year program, taking courses that run the gamut from mechanics, hydraulics, and electronics to turbine siting, worker safety, and data acquisition. With an associate's degree in applied science, they emerge primed to enter a job market where they find themselves competing with a growing cohort of well-trained technicians, many of them coming out of similar programs at places like Iowa Lakes Community College.

You can tell that the wind industry has penetrated the national ethos when a community college's training program for wind technicians is featured on a prime-time TV ad. That is exactly what happened when a commercial for Duracell batteries showed young guys in hard hats climbing a turbine and inspecting circuits, with the voice-over saying: "At Iowa Lakes Community College, the students learn to keep America's wind turbines going. And to keep them safe, the only battery they trust in their high-voltage meters are Duracell rechargeables."[24]

The Wind Energy and Turbine Technology Program at Iowa Lakes is housed in an attractive modern building on the college's campus in Estherville, a small village in the northwest corner of the state. This is a far cry from Cloud County's temporary rented quarters, shoehorned between the Dollar General store and the Dragon House Chinese restaurant in a Concordia strip mall. Ahmad Hemami, an instructor in the fundamentals of electricity, guides me through a half-dozen well-appointed teaching labs and a nearly equal number of classrooms. In the labs, students gather around electronic control boards, computer monitors, and wind-measuring anemometers, looking up only briefly as we walk through.

Ahmad then takes me to see a Vestas V90 nacelle, neatly mounted as a floating classroom in a courtyard just off the main corridor. I climb a single short ladder leading up to the suspended capsule and finally get to eyeball what I would have seen if I'd completed my climb at Meridian Way. The nacelle's cramped metallic interior feels like a cross between a submarine engine room and a space shuttle cabin. To make room for students, there is no main shaft running the length of the chamber and other large components, such as the gearbox and generator, have also been removed. Even so, the nacelle is a crowded maze of circuit boxes, electric motors, hydraulic hoses, and cooling fans to keep workers and machinery within a bearable temperature range. It's not hard to imagine how tightly packed a nacelle must be when filled with all the machinery that's needed to turn motion into power.

Iowa Lakes may be in a remote corner of a rural state, but from a wind technology perspective, it's in an ideal spot. Aside from having its own fully functioning wind turbine just a half-mile from campus, the college sits in a region that is replete with commercial-scale wind farms. "They're right smack-dab in northwest Iowa, where we need all the technicians," says Clipper's Bob Loyd, who sits on the Iowa Lakes board of advisers and is an enthusiastic booster of the program. The college also enjoys easy access to technology leaders at larger educational institutions like the University of Iowa and Iowa State. Barry Butler, dean of the University of Iowa's College of Engineering, is unstinting in his praise for the program and is in awe of the program's physical plant: "The laboratories they have are just enviable, even by our standards," he says. "I'd love to have some of the equipment they have."[25]

Among the students at Iowa Lakes are seasoned tradespeople like Joe Brightwell and Richard Dunham. Joe was a unionized electrician in Montague, Michigan, until his work dried up with the collapse of the housing industry.[26] Richard was a general contractor in Atlanta. When the recession devastated his business, Richard was already in his early fifties, yet he coolly set about searching for a new career. As a first step, he decided that he wanted to work in a field that would draw on his university training long ago in mechanical engineering. Next, he surveyed different energy technologies, ruling out several of them as less-than-promising employment prospects. "I wasn't convinced that nuclear power was going to come back, and there's already a lot of people in coal," he reasons. "So I looked at the alternative energies. Of course, wind seemed to be the most viable, and I knew it was something I would be interested in because there's a lot of mechanics as well as electrical work."

Though sold on wind, Richard is far from a true believer when it comes to climate change. "People are being persuaded falsely that there is a traumatic experience that's going to happen in the next fifty to a hundred years, like glacier melting and so forth," he says with obvious derision. On the other hand, he disdains coal company profiteering, loathes American reliance on Middle East oil, and believes that tapping the wind is an important way to secure American energy independence.

While students like Joe and Richard have already logged in decades in the technical trades, most Iowa Lakes trainees are young, barely out of high school. Paul Johnson is one of them, coming to Iowa Lakes after a short stint at a small college in North Dakota. The valedictorian at his hometown high school in northern Minnesota, Paul has a little trouble squaring his current studies with the expectations built up over his high school years. "In high school, these technical programs were kind of frowned upon," he acknowledges. "It was expected you could do better." He tells me that he hasn't managed to get any of his high school classmates excited about wind power, yet he speaks with animation about the industry's value in creating well-paying rural jobs. Growing up on a small farm with two parents who needed outside jobs to make ends meet, he is painfully aware of the strains facing family farming and has a sober respect for the difficulties in finding new local jobs to fill the vacuum.

Wind developers make a point of hiring local labor wherever possible. They know that when county commissions and economic development agencies weigh the pros and cons of inviting a new wind farm into their communities, the promise of new jobs will be a major consideration. Construction crews like the one led by Steve Maples at Meadow Lake have large contingents of workers hired through local unions. O&M teams like Leo Jessen's at Grand Ridge also draw from neighboring areas; Leo himself grew up and now lives in the town of Ottawa, Illinois, a dozen miles from the wind farm field office.

While many of the workers on wind farm construction crews and O&M teams are local, few of them are women. Project development teams often include women in supervisory and administrative roles; Carole Engelder is one example. But in the field, they are a rarity. At Meadow Lake, there were none in Steve Maples's ninety-man-strong construction crew. At Grand Ridge, I observed another construction team that included just one woman slogging through puddles on a rainy spring day, brushing and hosing mud off tower sections that had been lying in the fields. Meanwhile, the men busied themselves with their big machines, hoisting the freshly cleaned steel tubes into place.

On the operations side, Michelle Graham is an interesting and unusual hybrid. On one hand, she juggles a range of roles in the Meridian Way field office. She prepares daily reports on turbine availability, wind conditions, and needed repairs; she leads informational tours of the wind farm; and she maintains constant ties with the wind farm's participants and abutters, making sure they aren't caught off guard when an O&M crew goes out to repair a blade or replace a gearbox.

At the same time, Michelle has worked hard at defining a useful role for herself up-tower. To do this, she first had to get over her aversion to heights, so she began practice-climbing up a twenty-foot ladder that her husband Bruce—a wind instructor at home as well as at Cloud County Community College—had lashed to the side of their barn. "I strapped on his climbing harness and climbed up and down that ladder fourteen times," she told me. Afterward, she collapsed on the couch, her body exhausted and her hands blistered. That was in July 2009. By December, she had reached the top of a Vestas V90, where she spent several hours learning how to audit the services performed by O&M technicians—making sure bolts were tight, oil and grease levels were sufficient, and conditions were clean. Today, with her fears behind her, Michelle views up-tower quality assurance auditing as part of her on-the-job repertoire.[27]

Rare among workers in the field, women are nearly absent from the student rolls at technical training programs. At Cloud County Community College, only four out of a hundred students in the 2010–11 academic year were women.[28] The previous year, Iowa Lakes had just five women enrolled in its program.[29]

One of them was Loma Roggenkamp, a native Pennsylvanian who came to the wind energy field after trying her hand at marine biology in Maine. "I thought the ocean was our final frontier, and we were going to find things to save the world out there." Unable to land a job that drew on her studies, she worked in a graphics firm for several years. Then one day her mother spotted a wind technology ad on the Internet, and Loma immediately went online to explore possible points of entry. Training would be needed, she knew, and Iowa Lakes kept coming up as a great place to gear up for a career change. Soon she enrolled in the college with her family's blessing.

Immediately on graduating from Iowa Lakes in June 2010, Loma was snapped up by Siemens Energy to work as an O&M technician. First she was dispatched to Houston for three weeks of training; then she was sent to her "home farm," a 200-megawatt facility in Glenrock, Wyoming, 5,000 feet above sea level. At Top of the World, as the wind farm is fittingly called, Siemens operates forty-four 2.3-megawatt machines—just over half the farm's installed capacity.

In a phone call at the end of her first day up-tower, Loma's exhilaration is palpable. An athlete, she relishes the climb—nearly 300 feet straight up. "I can climb every day, and I don't have to pay for a gym membership!" she quips. And then there's the view. The Siemens nacelle opens like a clamshell, giving workers a stunning vista onto the surrounding countryside. "I can troubleshoot a turbine and look at nature at the same time," she says.

"I'm making a difference and I'm needed in this industry," Loma tells me with confidence and pride. That difference isn't just advancing a technology she believes in; it's helping women find safe and welcoming pathways into the wind energy workforce. Not surprisingly, much of the safety training for O&M technicians is focused on men; she wants to help develop training guidance for women. "Men aren't the only ones with appendages that are vulnerable," she says. Along with opting for the greater comfort of a safety harness configuration shaped like an "X" rather than an "H" in the chest area, women need to be cautioned not to wear flammable synthetic undergarments that can melt and adhere to the skin in case of fire.[30]

Loma is now working with a nonprofit group, Women of Wind Energy, to develop these training materials. She has found an enthusiastic ally and supporter in Kristen Graf, the organization's executive director. Kristen attributes the dearth of women in technical jobs partially to the stereotype that these sorts of positions are traditionally the domain of men. Women don't necessarily hear about wind energy training and hiring opportunities, so opening up the flow of information about schools and jobs is one needed step. Even for those women who stand up to the stereotypes, surmount the often-subtle biases, and get hired, Kristen says that working conditions built on a male culture can sometimes be unwelcoming and isolating. "Some women are good at playing the role of being 'one of the guys,'" she observes, "but that shouldn't be a requirement if you have all the skills needed for a particular job." Her goal, and that of her organization, is "to be more proactive in creating a working environment that fits everyone."

At least as troubling to Kristen is the underrepresentation of women in higher-level corporate positions. Women of Wind Energy, which has thirty state chapters and a few additional branches in Canada, provides fellowships for women to attend the AWEA annual meeting, a multiday conference and trade fair that draws tens of thousands of participants each year. Newcomers to the job market get easy access to wind energy employers at these gatherings. Kristen describes Denise Bode, CEO of AWEA, as "a real champion" in raising the profile of women in the industry.[31] Even with those efforts, there remains a very large gender gap in wind energy management—a gap that is reflected by the composition of the AWEA board: out of twenty-eight members serving as of April 2011, only five were women.[32]

Jeanna Walters, a Cloud County graduate who has gone on to pursue a bachelor's degree in environmental science at Kansas State, sees the challenge facing women in wind as beginning with the school system. "We really need to push the science and math on girls in high school," she insists. Even if women don't want to end up working in construction or climbing turbines, she is sure that opportunities abound for women in engineering, land acquisition, development oversight, financing, operations management, and marketing. "This is definitely an industry that girls can be involved in—and should be."[33]

It's hard to predict how many U.S. jobs a thriving wind energy sector might provide in future years, but that hasn't kept the U.S. Department of Energy (DOE) from trying. In a report called *Wind Power in America's Future: 20% Wind Energy by 2030,* it looked at the prospects for generating electricity from wind. If wind were to supply 20 percent of America's power by 2030, the DOE research team estimated that annual installations of new wind-generating capacity would have to reach 16 gigawatts by 2018—an increase, but certainly not an impossible leap from an installation level that came very close to 10 gigawatts in 2009. Under this scenario, total installed wind power nationwide would reach 300 gigawatts by 2030—sufficient to generate a fifth of the 5.8 billion megawatt-hours projected to be the overall U.S. demand for electricity by that date. This is *not* a formula that optimistically flatlines U.S. electricity use between now and the target date; to the contrary, a 39 percent increase above total consumption in 2005 is built into the calculation.[34] If we actually became a nation that valued energy conservation more than we do today, 300 gigawatts of installed wind power could end up providing well over 20 percent of the nation's power needs by 2030.

Under the *20% Wind Energy by 2030* scenario, manufacturing jobs directly related to producing wind turbine components and subcomponents would top 30,000 by 2021, peaking at 32,835 in 2028. While factory work would somewhat slacken thereafter, ongoing expansion in generating capacity—both onshore and offshore—and the need to re-power aging wind plants would guarantee a continued high level of employment in the manufacturing sector. In construction, jobs would average over 70,000 a year from 2019 through 2030. And in wind farm operations, total jobs would reach 76,667 by 2030—about 28,000 in on-site O&M and another 48,000 in utility services and subcontractors. Adding them all up, DOE foresees about 180,000 jobs directly linked to wind energy as the 2030 target date approaches.

Beyond all the "direct" jobs in the wind energy economy, DOE also explores the "indirect" employment benefits of growing this sector. These jobs include the producers and suppliers of steel, fiberglass, and other materials that are used to build wind turbines; the companies that produce the parts that go into turbine components and subcomponents; and the providers of banking, accounting, legal, and other services to wind turbine manufacturers and wind farm contractors. These indirect jobs are expected to number about 100,000 annually in the years leading up to the 2030 target date.

Finally, DOE draws an even wider circle around the "induced" jobs resulting from consumer spending by people directly and indirectly employed in the wind energy sector. A Clipper factory worker buys a new pair of jeans in a local store; an O&M technician takes his family out to dinner; a crane operator stays at a local motel. The DOE team attributes another 200,000 jobs per year to these induced economic activities.[35]

Folding induced jobs into the assessment of wind energy benefits may go farther down the speculative road than some are ready to travel. But even setting that outer circle of employment impacts aside, we are looking at a roster that rises to more than a quarter-million direct and indirect jobs if we pursue DOE's *20% by 2030* goal. A technology commitment that advances America's energy independence and reduces our nation's carbon footprint while creating hundreds of thousands of new, skill-based jobs—isn't this a path worth taking?

CHAPTER SIX

The Path to Cleaner Energy

BEYOND BRINGING JOBS TO tens and eventually hundreds of thousands of Americans, can we expect wind energy to make a real difference in the way we fuel our economy? How much can we count on wind energy to rein in the pollution that is compromising our health and warming the Earth's atmosphere? And what will it cost us to make wind a major part of our energy diet?

Looking at wind energy's contribution today, it may seem premature—even presumptuous—to think of the technology as a game-changer. During 2010, wind supplied 2.9 percent of America's power needs.[1] Coal, during the same period, delivered 45 percent of our electricity, and natural gas generated another 24 percent, followed closely by nuclear power, with a 20 percent share of overall output. Another 6 percent of our electricity came from conventional hydroelectric dams, while solar energy—photovoltaics and thermal systems combined—barely registered a blip on the screen at 0.03 percent.[2]

From the broader vantage point of energy use across all sectors, wind's current status looks humbler still. Electricity generation amounts to a little less than 40 percent of total U.S. energy consumption, with most of the rest relying on the direct burning of fossil fuels to run our vehicles, stoke our industries, and heat our buildings. So that means wind meets slightly more than 1 percent of our overall energy needs.

Wind energy's role in weaning America off fossil fuels may be modest today, but its untapped potential is vast. In February 2010, the government-run National Renewable Energy Laboratory (NREL) released the results of a mapping effort that gauges the full magnitude of land-based wind energy as a resource that can serve America's power needs (see table 1). First NREL identified "windy land areas," which it defined as areas where the average winds are strong enough, at 80 meters (262 feet) above ground, to allow turbines to produce electricity at a minimum of 30 percent of their full installed or "nameplate" capacity, averaged annually. (A 30 percent "capacity factor" is considered moderately robust.)[3] Areas with annual wind speeds averaging at least 14.6 miles per hour were regarded as meeting this threshold.

From this gross measure, NREL then subtracted territory that it deemed unlikely to be developed for wind because of conflicting uses or characteristics—urban areas, parks, designated wilderness, and areas with water features that could hinder wind development. In my home state of Massachusetts, NREL rated less than 1 percent of the state as sufficiently windy, and it disqualified 88 percent of that small area because of conflicts. At the other end of the spectrum is Kansas, where almost 90 percent of the land meets the threshold for windiness, and only 10 percent of that windy area has been sidelined by NREL because of conflicts.

I remember loving those spin art kits that were toy store staples in the 1960s. After dribbling paints from ketchup-like dispensers onto a rectangular sheet of cardboard, I'd flip the switch of a battery-operated spinning wheel. Within seconds, a swirling, sometimes lurid maelstrom of colors emerged. On first glance, NREL's digitized wind map of the United States reminds me of those creations. In the Southeast, spring green splashes across the Atlantic coastal states, reaching as far inland as Tennessee, Alabama, and eastern Mississippi. This color connotes average wind speeds of less than 11.2 miles per hour—well below what's needed to make wind a strong competitor with other power-generating fuels. Moving up into the Northeast, green yields to splotches of yellow and shades of brown, suggesting wind speeds edging up to 15 miles per hour. The same dappling of forest colors spreads from the Pacific Coast through Idaho, Utah, Nevada, and New Mexico. Then the real excitement begins. From the Rockies, across the Great Plains, stretching from eastern Montana and the Dakotas down to Texas, deep veins run from rust to violet to purple, plunging occasionally to midnight blue, revealing vast areas where average wind speeds range from 15 to 20 miles per hour.[4] These expanses are the stuff of wind developers' dreams.

Adding up all the qualifying windy areas, NREL projects that America could install up to 11,000 gigawatts of land-based wind power, yielding roughly 38.5 million gigawatt hours of electricity per year. *That's nine times our total electricity production today!*

Offshore wind is another huge energy resource, virtually untapped in America, though a growing contributor to European power generation.[5] NREL has concluded that U.S. ocean waters within fifty nautical miles of land, plus U.S. portions of the Great Lakes, could yield 4,000 gigawatts of wind-generating capacity,[6] bringing our combined onshore and offshore wind energy potential to 15,000 gigawatts. *That's nearly fifty times the amount of wind energy that the Department of Energy says we will need to harness to supply 20 percent of our nation's power needs by 2030.*

A whole host of environmental, aesthetic, and logistical concerns can make many a wind-rich site a poor prospect for building a wind farm. From my own advocacy for onshore and offshore wind energy in New England, I know how huge a leap it can be from identifying a resource to seeing wind blades spinning on the horizon. Whether the concerns are about harm to wildlife, noise, inadequate access to transmission, or simply the interrupted view, the realm of sites acceptable to policymakers and the public is considerably smaller than the universe of wind-worthy areas. We may not want to develop every promising wind site in the nation, but it's clear that wind energy could become a mainstay of our energy economy using just a small fraction of the available resource.

We may have more wind on hand than we can possibly use, but what will we gain by tapping this resource? One clear "win" is the opportunity to reduce America's contribution to global greenhouse gas emissions.

The carbon emissions gap between wind power and conventional fossil fuels is nothing short of stupendous. The Swedish utility Vattenfall, Europe's fifth-largest electricity producer, has prepared a life-cycle environmental assessment of a range of generating sources. Typical of cradle-to-grave environmental assessments, this evaluation looks at the materials used and the pollutants generated at all stages of a technology's implementation. For coal, this includes building the power plant; mining, purifying, transporting, and storing the coal; burning it to produce electricity; disposing of coal ash; and decommissioning the plant at the end of its productive lifespan. For natural gas, the study's scope is similar, beginning with drilling and running all the way through power plant decommissioning. For wind, the coverage extends from manufacturing and operating the turbines to end-of-life dismantling. Taking all these factors into account, Vattenfall reports that wind, averaged over its full life cycle, produces about 10 grams of carbon dioxide (CO_2) per kilowatt hour of electricity generated, while coal releases 600 to 700 grams per kilowatt hour—that's sixty to seventy times as much CO_2 per unit of power. Gas, depending on the technology used, ranges from 400 to 1,300 grams of CO_2 per kilowatt hour—40 to 130 times as much CO_2 as wind.[7]

Here in the United States, NREL conducted its own life-cycle assessment of coal-based power production in 1999. It found that the average U.S. coal plant emitted 1,022 grams of CO_2 per kilowatt hour, and newer coal-burning facilities released 941 grams per kilowatt hour.[8] In more recent years, the performance of our coal plants hasn't improved very much. In 2008, the average coal-burning facility put out about 995 grams of CO_2 per kilowatt hour. Newer, more efficient plants performed only slightly better, with the hundred best-performing units averaging about 865 grams per kilowatt hour.[9]

Vestas has been an industry leader in preparing and publishing life-cycle analyses of its wind turbines, following guidelines established by the International Organization for Standardization (ISO). Looking at the V90 3-megawatt machine that stumped my climbing efforts at Meridian Way, the company estimates that each turbine will generate about 158,000 megawatt hours of electricity over the course of its twenty-year lifetime. By using wind rather than coal to generate this power, Vestas expects that a single turbine will spare the environment about 130,000 tons of CO_2 pollution over the course of its useful life.[10]

Another tool for evaluating different power-generating technologies is called an "energy balance." In an energy balance, the energy consumed by a given technology is compared to the energy it produces over its full operational lifetime. Vestas analyzed the V90 in this manner. First it looked at the extraction and production of raw material inputs like steel. Next it evaluated the energy used in manufacturing, installing, operating, and ultimately decommissioning the turbine. (The steel used in the tower and other major components can be recycled once a turbine is taken down, by the way.) Balancing all those inputs against the turbine's estimated lifetime energy output, it found that each V90 produces enough electricity to cover its full lifecycle energy inputs within the first 6.6 months of operation.[11] From that point onward, all of the turbine's generated power is essentially carbon-free.

For coal, the story is very different. For every moment a coal plant operates, from its first day in service to its last, huge quantities of energy are consumed and correspondingly huge amounts of carbon are released into the air. Giant excavators keep grinding away at coal seams, diesel-powered trains continue to pull endless chains of coal cars across hundreds of miles of open country, and power plant boilers burn the fuel in nearly unfathomable quantities right up until the day the plant shuts down. The energy balance for coal and every other fossil fuel begins in a deficit that only deepens with every new ton of fuel consumed.

The Department of Energy has looked at the CO_2 emissions that we could avoid by generating 20 percent of our power from wind. It estimates that this switch to wind power would cut our yearly CO_2 emissions by 825 million metric tons, or about 26 percent of all projected CO_2 emissions from the power sector in 2030.[12]

In addition to burning finite energy resources throughout their years of operation, coal and other fossil fuel–based power plants spew massive amounts of dangerous pollutants into the atmosphere. Sulfur dioxide emissions are about 48 times higher and nitrogen oxide emissions are 42 times higher for coal plants than for wind farms, averaged over the full life cycle of each facility. (For wind farms, air emissions—like the energy inputs already discussed—are primarily associated with the fuels burned in producing, erecting, and dismantling the turbines, not with day-to-day operations.) Methane, an ozone precursor and a greenhouse gas that—gram for gram—is much more potent than CO_2, is produced at levels 320 times higher for coal than for wind. Modern, combined-cycle gas facilities fare better, although the gap is still huge: sulfur dioxide levels are 4.6 times larger, nitrogen oxides are 11.4 times greater, and methane emissions are 124 times as high as wind.[13] Adding this heavy dose of pollutants to the CO_2 loadings caused by coal and gas facilities, it's clear that our ongoing reliance on fossil fuels comes at a very high environmental price.

The environmental drawbacks to fossil fuels, grave though they are, pale by comparison to the "perils of the peaceful atom."[14] In one of the most unsettling paradoxes of modern times, mounting concern about global warming has brought nuclear power to the fore as a "carbon-neutral" technology that is ostensibly friendlier to the global climate than coal or other fossil fuels. A few veteran voices in the environmental community, like Stewart Brand, founder of *Whole Earth Catalog* and an early mentor of California governor Jerry Brown, have even joined this chorus.[15]

When I was in graduate school in the early 1980s, I had a professor who was fond of explaining statistical probability with the reminder: "Rare events *do* happen." Those words haunted me at the time. Only a few years had passed since a combination of faulty design and human error had led to the partial meltdown of a nuclear reactor at Three Mile Island, near Harrisburg, Pennsylvania. Although no public health disaster ensued, the accident caused the American public to doubt the blithe assurances of nuclear industry leaders about the safety of their technology. Then Chernobyl melted down in 1986, and the nuclear nightmare that had been averted at Three Mile Island became a terrifying reality. Tens of thousands were exposed to dangerous levels of radiation, thyroid cancer began appearing at an alarming rate, and about 9 percent of the Ukraine and 23 percent of Belarus remain contaminated today.[16]

The disaster at Japan's Fukushima Daiichi nuclear complex came as the next tragic reminder of nuclear power's menace. As I learned about the many tens of thousands of people who had to flee their homes and the hundreds of emergency workers who subjected themselves to extremely hazardous radiation levels while struggling to contain the damage, my thoughts kept circling back to my professor's warning about rare events. Should nuclear safety experts really have to plan for a double blow as improbable as an earthquake of practically unprecedented magnitude and a tidal wave of epic proportions hitting the same reactor complex in rapid succession? And yet this is exactly what happened within a few decades of the plant's construction. It's too soon to know the full impacts of this catastrophe, but the folly of relying on nuclear power to meet the twenty-first century's energy needs couldn't be clearer.

As horrific as accidental reactor meltdowns can be, nuclear power's hazards to humanity and the environment extend beyond the realm of the unintended. In the post-9/11 era, we already know the havoc that can be created by commercial jetliners commandeered as guided missiles. In fact, targeting a nuclear reactor was an option considered by Al Qaeda in the period leading up to the World Trade Center and Pentagon attacks.[17] While nuclear industry spokespeople have claimed that a plane's likelihood of breaking through a reactor's concrete containment and exposing the reactor core is low, the National Academy of Sciences has warned that an aerial attack on a power plant's spent fuel storage pool could drain the pool and trigger a high-temperature fire, releasing potentially deadly quantities of radioactivity into the environment.[18]

Short of a direct attack placing nearby populations at risk, it would be foolhardy to ignore the hazards of further proliferating nuclear fissile materials in an unstable world where rogue regimes and terrorist groups are hustling to develop their own nuclear stockpiles. Those who admire France's commitment to nuclear power often gloss over the huge risks involved in that country's reliance on fuel reprocessing to keep its reactors going. Separated plutonium—a key product of reprocessing—is a prime feedstock for nuclear weapons, much more valuable to potential bomb makers than decaying fuel rods. This danger has led Peter Bradford, former member of the U.S. Nuclear Regulatory Commission, to warn: "In a world in which reprocessing becomes an integral part of the fuel cycle, the International Atomic Energy Agency's proliferation safeguard system would face challenges that it is not capable of dealing with because separated plutonium is so much easier and quicker to transfer into a weapon than . . . fuel rods that are in a reactor pool."[19]

Then there is the nuclear waste. Once the fuel rods reach the end of their useful lives, they are removed from the reactor core and are placed in temporary storage, usually at the reactor site but sometimes off-site, waiting for a safe method of long-term disposal to be discovered. In nuclear power parlance, "long-term" means tens and even hundreds of thousands of years for certain radioactive isotopes that are waste products of the fuel cycle.[20] Eventually, when a reactor is decommissioned, much of the plant's hardware must also be sequestered until its radioactivity subsides. Because of the costs and dangers involved in dismantling them, only three of the twenty-four permanently closed U.S. power reactors have been fully decontaminated; nearly half are in a holding pattern that the Nuclear Regulatory Commission calls SAFSTOR, deferring decontamination to a later date.[21]

Ever since the early 1980s, debate among experts and politicians has raged about the viability of a deep underground high-level nuclear waste repository at Yucca Mountain in Nevada. By the time President Obama delivered on a campaign promise to halt further work on the Yucca project in 2009, the government had already spent $7.7 billion at the site. Over its full operating lifetime, the facility would have cost taxpayers and electricity ratepayers close to $100 billion.

With Yucca on hold indefinitely, the search for a viable long-term storage solution for nuclear waste continues. Meanwhile, every operating reactor continues to produce about 20 metric tons of used nuclear fuel per year, filling up temporary storage facilities way beyond their planned capacities. Much of the waste is stored in pools just like the ones that erupted into steaming, highly radioactive cauldrons at Fukushima Daiichi. "Hubris" doesn't begin to plumb the depths of our imprudence in perpetuating and possibly expanding our reliance on an ultra-hazardous technology that we have no idea how to manage into the indefinite future.

Even the carbon neutrality claim about nuclear power is notably weak. Like fossil-fuel power plants, nuclear reactors demand an ongoing supply of fresh fuel, in the form of highly enriched uranium. Each stage in the nuclear fuel cycle consumes energy—from the equipment used to mine uranium, to the power demanded by the various stages of concentration, enrichment, and fuel-rod fabrication, to the transportation of nuclear materials from one stage to the next. And what about the massive amounts of carbon emitted by the huge mobilizations involved in containing and cleaning up Three Mile Island, Chernobyl, and Fukushima?

Beyond wind energy's clear merits as a safe, low-carbon power producer, it offers some dramatic new ways to lessen our dependence on foreign energy resources. The opportunities begin with oil.

Wind energy opponents often argue that harvesting the wind will do little to reduce America's reliance on Middle East oil. They rightly point out that a very small fraction of America's electricity comes from oil-fired power plants, most of which were phased down, if not out, in the 1970s, when skyrocketing oil prices in the post–oil embargo years triggered a massive switch to domestic coal. Evidence of our success in cutting back on the use of oil for electricity production is unequivocal: oil supplies less than 1 percent of our present-day power needs, way down from 16 percent in 1973.[22] If only we had been remotely as effective in trimming our overall appetite for oil during this period! Total U.S. oil consumption is nearly 2 million barrels a day higher today than it was in 1973.[23]

Wind power's advantage may not be in displacing oil-based power generation, but it could make a different kind of dent in our oil dependence by changing the way we move people and goods around the country. Transportation ranks second only to power generation as a U.S. energy user, accounting for 27 percent of all energy we consume. And of the energy we use for transportation, almost all of it—97 percent—comes from oil.[24] Anything we can do to trim our oil use in this sector will go a long way toward curbing America's greenhouse gas emissions. At the same time, the security gains could be enormous. The turmoil now sweeping North Africa and the Middle East only magnifies the madness of our reliance on foreign nations for so much of our oil supplies. Wouldn't it be a relief if our transportation lifeline led to wind farms in Kansas and Wyoming rather than oil fields in Iraq, Libya, Saudi Arabia, and the United Arab Emirates? By matching up large networks of electric cars with a robust, wind-supplied grid, this transformation is within our reach.

Late at night, with dinner already prepared, the dishes washed, and televisions and computers shut off, the wind still blows. At such times, turbines can be taken off-line to avoid overloading the grid with excess electrons, but this wastes a fuel source that is essentially free. The capital costs of building a wind farm are considerable, as they are for conventional coal, gas, or nuclear power plants, but what distinguishes wind from those other power sources is the absence of any ongoing fuel cost. Increasingly, energy planners and innovators have been coming up with new ways to make fuller use of wind-generated power at night and other times when routine demand for electricity is low. Creating centrally managed networks of plug-in electric vehicles is one such solution that Denmark is now pioneering.

Knud Pedersen is vice president for research and development at Denmark's utility conglomerate Danish Oil and Natural Gas (DONG). As its full name implies, DONG was once a company that focused largely on fossil fuel production and distribution. Today it is a leader in tapping wind and other forms of renewable energy. When we met in his glass-walled office overlooking one of Copenhagen's busier traffic arteries, Pedersen spoke excitedly about the steps DONG is taking to reduce its carbon footprint. The largely state-held company owns over half of Denmark's conventional power-generating capacity, including several of the country's biggest coal-fired power plants. It also operates distribution lines serving nearly a third of the nation's electricity customers. "[E]ighty-five percent of our production capacity is fossil-based or CO_2-emitting," Pedersen told me. "Our goal is to turn that upside down by 2040." DONG is committed to cutting its CO_2 emissions per kilowatt hour in half by 2020, as a first step toward having 85 percent of its electricity come from wind, biomass-derived fuels, and other renewable resources twenty years later. This would place DONG substantially ahead of the European Union's new mandate calling for an 80 percent reduction in regionwide greenhouse gas emissions by 2050.

Already DONG has invested heavily in wind power at home and abroad. With well over 1,000 megawatts of installed wind capacity, the company distributes some of its wind-generated power in Denmark and markets the rest to other countries in northern Europe. Its plan is to own at least 3,000 megawatts of wind-based generation by 2020.[25] The more wind DONG acquires, the bigger will be the challenge of matching its customers' consumption patterns to the available supply of wind power, which varies with the strength of the winds.

This is where electric cars enter the picture. Together with Israeli innovator Shai Agassi and his Silicon Valley firm, Better Place, DONG is preparing a network of tens of thousands of electric vehicles whose owners will plug them in when they arrive home, leaving it to DONG to decide when it would be cheapest to charge them up for the next day's driving. Pedersen assured me that participants in this program will not be subjected to hardship conditions. "You should see the car as your normal car, with the same convenience apart from no noise—very speedy acceleration and very comfortable," he said, opening a brochure that features two of the models that will be used. One is a small SUV, the other is an attractive, mid-sized sedan; both will be manufactured under agreement with Renault-Nissan.[26]

In its early implementation, Better Place is looking to develop electric car networks in relatively confined geographical areas. This is important because, in addition to off-peak charging, the Better Place system requires battery swap stations to be built along highways so that cars traveling longer distances won't be caught without a charge. Small countries like Denmark and Israel are ideal, as travel distances tend to be relatively short. For the same reason, Better Place is focusing much of its initial attention in the United States on Hawaii, although the company is also working with state and local officials in California to create an electric car network in the Bay Area.

To avoid the cost and inconvenience of battery replacement stations, a number of U.S. researchers are thinking about building plug-in networks based on hybrid-electric cars and trucks instead of the all-electric vehicles proposed by Better Place. Under this approach, utilities would have the same leeway to select low-demand times for recharging batteries, but the use of hybrid engines would lift the limitation on driving distances.

It's no flight of fancy to imagine half of America's cars and small trucks running on electricity in a few decades' time. Just as we geared up to produce ever-greater quantities of internal combustion vehicles through most of the twentieth century, we could shift that vast and now faltering productive capacity toward a challenge truly worthy of our engineering creativity and competitive zeal. What better opportunity is there for the U.S. auto industry to recover from decades of short-sighted and financially calamitous decision making than by retooling to become a world leader in the manufacture of electric vehicles? If we don't seize this opportunity, it's clear that our European and Asian competitors will.

Looking to the future, America's vulnerability to foreign energy supplies may not be limited to oil. More than half of American households today are heated with natural gas, and more than a fifth of our electricity is gas-generated. While most of that gas still comes from within our borders, known and proven U.S supplies are limited, amounting to about 273 trillion cubic feet[27]—little more than eleven times the amount of gas that we now consume in a single year.[28]

American gas industry proponents are quick to point out that additional "unproved" but "technically recoverable" reserves may be as much as nine times that amount.[29] They are less ready to own up to what is involved in accessing those reserves.

Unconventional recovery methods have substantially expanded U.S. gas supplies in recent years. Shale formations in Appalachia, Texas, and the Rockies are prime sites for drilling operations that provide about a fifth of our natural gas supply—up from only 1 percent at the turn of the millennium. These operations use high-pressure water, converted chemically to a viscous gel, to fracture rock formations that are often a mile or more beneath the surface. Effective in releasing trapped gas, hydraulic fracturing ("fracking") is a huge drain on local water supplies, often using up to 3 million gallons of water per job. It also poses a pollution hazard whose dimensions are largely obscured by the companies' refusal to divulge the chemicals in their fracking compounds. Along with the fracking chemicals, leachates from shale rock enter the water stream, often polluting it with salt brines as well as radionuclides and heavy metals like arsenic and barium. Companies at times re-inject these fluids into shallow wells, but public concerns about drinking water contamination have brought this practice into serious question. The alternatives—discharging into surface waters or evaporation ponds—have their own environmental downsides.[30]

Coal-bed methane is another unconventional source of natural gas, produced most intensively in New Mexico, the Rocky Mountain states, and Appalachia. Coal seams tend to be closer to the surface than shale, but they have to be dewatered before methane can be extracted, yielding large volumes of water that is often highly saline. Discharging this water into surface streams, a common practice, alters natural habitats and also can endanger drinking water supplies.

Through deeper drilling, "re-fracking" shale gas wells that have already been tapped, and other technology advances, we may be able to access greater amounts of gas than are within our reach today. Yet gas industry insiders are already beginning to panic that their recent projections about future shale gas yields have relied on unduly optimistic assumptions about cost, well productivity, and technical feasibility.[31] This may shorten today's romance with cheap domestic gas, shifting American utilities and other consumers toward foreign gas suppliers long before our own reserves run out. And where are the world's biggest gas resources? In Russia, whose known reserves outstrip America's by a factor of six; in Iran and Qatar, each of which has more than three times the known reserves we have here in America; in Saudi Arabia, whose known reserves roughly match our own; and in places like Algeria, Iraq, Libya, and Venezuela, where sentiments toward our country range from volatile to overtly hostile.[32]

For those who doubt that we could sink into dependence on yet another imported fossil fuel, we need only look back at the decades that have passed since Richard Nixon and Jimmy Carter called upon lawmakers and the American public to rally behind an effort to free ourselves from the stranglehold of the Organization of Petroleum Exporting Countries (OPEC) in the 1970s. As we sat in those long fuel lines and watched gasoline prices skyrocket in the months following the 1973 oil embargo, who would have believed that we would allow our reliance on foreign oil to rise from 28 percent then to 60 percent today?

Wind energy won't eliminate our appetite for natural gas, just as it won't free us from our hunger for imported oil. It can, however, help keep our natural gas consumption within bounds. The Department of Energy predicts that we can reduce our reliance on natural gas–fired power plants by half if we boost our wind energy production to 20 percent by 2030. This, DOE says, would reduce overall U.S. demand for natural gas by 11 percent.[33]

We could curb our natural gas use further if Americans who now heat their homes with natural gas converted to electric heat pumps or other electric systems reliant on renewable energy–based power. The American Gas Association reports that more than 64 million U.S. homes are heated with gas, an increase of 27 million customers since 1970. Many of these gas users had previously relied on oil to heat their homes; their switch to gas was largely in response to the rising price of home heating oil during the years following the OPEC oil embargo. Imagine what would happen to our natural gas demand if, over the *next* forty years, we saw an equivalent shift in customers from natural gas to electric heat substantially generated by wind? And what if the 8 million households that still heat with oil switched over to renewable energy–based electricity? The reduced fuel demand and lowered greenhouse gas emissions could be enormous.

There is no question that shifting tens of millions of cars and home-heating systems to electricity will demand a much more robust U.S. electric sector than we have today. For this to help rather than further harm the environment, we will need to ensure that our expanded generation is based on renewable energy. Meeting this demand with new coal-fired or nuclear power plants would only replace the petroleum used in cars and the gas used for home heating with other fuel choices that pose their own very grave environmental and security hazards. We will also have to build a much more robust and sophisticated transmission network that is capable of carrying large quantities of wind-generated power from remote locations, both onshore and offshore, to major population centers. Jon Wellinghoff, chairman of the Federal Energy Regulatory Commission (FERC), sees both of these challenges as well within our reach. He envisions a "smart grid" with the capacity to manage a U.S. electric system that is 50 percent or more reliant on wind.[34]

Even short of fulfilling the FERC chairman's long-term vision, we will need to increase our investment in wind substantially if we are to meet the Department of Energy's goal of drawing 20 percent of our power from wind by 2030. On average, we must install 16 gigawatts of new wind power capacity every year between now and 2030 to meet the 20 percent target. While this is an ambitious goal, it would certainly be within reach if we adopted federal pricing policies that accounted for the full costs of our fuel choices: the global warming effects and air pollution damage to people, property, and natural resources caused by carbon-based fuels; the practically incalculable hazards of nuclear power; and even the costs of waging military campaigns to protect our energy interests abroad. No such pricing is on the immediate political horizon, but lesser steps are certainly within reach. The extension of federal production and investment tax credits for renewable energy projects is one such measure. Adoption of a federal renewable electricity standard is another, requiring all states to ensure that the utilities under their jurisdiction draw a substantial portion of their power from wind and other forms of renewable energy.

Happily, we needn't simply wait for the federal government to take action; more than thirty states have already adopted their own renewable electricity standards. Illinois, for example, has mandated that wind must provide three-quarters of its 25 percent renewable electricity quota for the year 2025, with three-quarters of that power mandated to come from wind.[35] In Kansas, utilities must acquire 20 percent of their electricity from renewable energy sources by 2020.[36] The nation's most ambitious renewable electricity standard, in California, requires utilities to draw 33 percent of their power from renewable sources by 2020. That's in addition to the abundant power that already comes from hydroelectricity—about 15 percent of the state's overall power supply. Qualifying renewable resources—mainly wind, geothermal, solar, and biomass—supplied 18 percent of the power used by California's three big utilities in 2010, falling just short of the state's previous renewable electricity mandate, which set a 20 percent renewables target for that year.[37]

In addition, eight states have passed laws requiring utilities to offer customers a green power option—the choice of purchasing some or all of their power from renewable sources. And in forty-two states, net-metering laws obligate utilities to buy excess power generated by customers who have installed wind turbines, solar panels, or other renewable energy–based power-generating systems.[38] These measures, together with a range of state renewable energy grant and loan programs, have tilled fertile ground for wind energy development across much of the nation.

~

Wind energy's price relative to other power-generating technologies is critical to determining how big a role it will play in meeting America's electricity needs. The Lawrence Berkeley National Laboratory has gathered wholesale price data on 180 wind projects installed between 1998 and 2009, representing 38 percent of all installed wind power during that period.[39] Through most of those years, wind power was at the low end of wholesale electricity prices, making it highly competitive with electricity coming from other energy plants. However, this relatively favorable relationship eroded in 2009, when a precipitous drop in natural gas prices—largely the result of expanded shale gas development—and a reduction in power demand as a result of the economic recession caused wholesale power prices to decline. By 2010, new wind projects commanded an average wholesale price of roughly $70 per megawatt hour, at a time when the wholesale power market averaged between $30 and $50 per megawatt hour.[40]

Despite this drop in gas prices, utilities have an ongoing stake in buying into the wind market, partially as a long-term hedge against the price instability of fossil fuels. Natural gas may be cheap today, but its price history has been a rollercoaster ride, and those swings very likely will continue in the future.[41] Coal prices, too, may be hard to predict with any certainty if mine safety and environmental protection measures are made more stringent in the years ahead. Utilities also value wind as a way to diversify their portfolios in anticipation of eventual federal rules that make fossil fuels more costly or require a reduction in power plant carbon emissions.

Utilities have a further—and very important—reason to invest in wind: they need it to meet their obligations under renewable electricity standards in the states that have adopted them. Take California as an example. Between 2002 and 2010, wind delivered about 83 percent of the power that utilities bought to meet the state's 20-percent-by-2010 renewable energy mandate.[42] Looking ahead, California utilities will need substantial new increments of wind, generated in-state or purchased from its neighbors, to achieve the state's new commitment to deliver 33 percent of its electricity from renewable sources.

While these and other considerations, such as customer preference and broader public opinion, may cause utilities to favor wind over other energy sources, pricing wind power competitively remains crucial. One major way to keep prices within bounds is to contain the costs of building and operating wind farms. The financial advisory firm Lazard Ltd. has examined the twenty-year life-cycle costs of different types of power plants, "levelizing" them by looking at the same parameters for each energy type, including up-front capital costs, the fixed and variable costs associated with operations and maintenance, and fuel costs.[43] In 2010, it found the cost of new wind farms to range from $65 to $110 per megawatt hour, making them competitive with new coal plants ($69 to $152 per megawatt hour) as well as new combined cycle gas plants ($69 to $96 per megawatt hour). Solar was much more expensive, with photovoltaics costing $134 to $192 per megawatt hour and solar thermal plants (generating steam from concentrated solar heat) costing $119 to $194 per megawatt hour. Nuclear came in at $77 to $114 per megawatt hour, though experts assume that, post-Fukushima, its cost will rise substantially given the increased safety measures that will be applied to any new plant proposal.

These numbers highlight the importance of the federal production tax credit for wind. Without it, wind energy's cost as calculated by Lazard would rise by $20 per megawatt hour, making it much less competitive with coal and gas.

Far and away the biggest factor affecting wind energy's cost is the price paid for turbines. Up-front capital outlays for equipment, construction, and financing account for 76 percent of all life-cycle costs associated with wind energy, and the turbine amounts to 75 percent of that initial investment.[44] There's good news in this regard, stemming largely from the growing competition among turbine manufacturers discussed in earlier chapters. From a peak in 2007–2008, just when Horizon was building the Meridian Way Wind Farm, turbine prices have been steadily declining. By late 2010, the average contract price for turbines had dropped nearly 20 percent,[45] and prices have continued their decline more recently.

Along with spending less on turbines, enhancing their performance is another way to get more from a given wind energy investment. Taller turbines and longer blades are among the changes that have had the greatest impact. Taller turbines generally produce more power because wind speeds at higher altitudes tend to be stronger, more constant, and less disrupted by near-ground turbulence, or wind shear, caused by landscape features like trees, buildings, and variations in the natural topography. This height advantage is magnified by the "cube rule" of wind power, establishing that the power available from the wind is the cube of its velocity. Given this relationship, it's clear that small increments of wind speed—easily achieved by raising tower heights—can yield big productivity gains. Increasing the average wind speed from 10 to 12 miles per hour boosts the wind power reaching a turbine by 73 percent; raising it to 14 miles per hour augments the available power by 175 percent.

Longer blades further enhance the power-generating potential of turbines. The power available to a turbine is directly related to the "swept area" of the rotor, determined by the familiar formula for calculating the area of a circle: $A = \pi r^2$. Applying this formula, we can see that small increases in blade length (that is, the radius) will translate into relatively large increases in a rotor's swept area, allowing much greater amounts of energy to be converted into electric power.[46] A 50-meter blade has a swept area more than half again as large as a 40-meter blade. The biggest rotor that Vestas makes today—120 meters in diameter—has a swept area 36 times as large as the 10-meter rotors that it was importing to California in the mid-1980s. The resulting power gains are impressive.

The first few decades of the wind energy era have seen dramatic jumps in tower height, from latticework structures a few dozen feet high in the 1980s to today's sleek tubular towers, with hub heights approaching and sometimes exceeding 300 feet. We have also seen turbine blades grow from slender slivers that you could load into the back of a pickup truck to multi-ton behemoths stretching half the length of a football field. From 1998–99 to 2009, the average hub height rose from 185 to 258 feet—an increase of 39 percent. The average rotor diameter expanded during the same period by 69 percent, from 159 feet to 268 feet.[47] These growth trends appear to be slowing for land-based turbines, but the current race to build higher-output offshore turbines is spurring the development of bigger-diameter rotors at sea.

On many levels, wind power is superior to the fossil and nuclear technologies that we rely upon so heavily today. It taps a resource that is abundant beyond our wildest dreams. It can be harvested without stripping the Earth of its mineral resources. Its daily operations neither pollute the local environment nor burden the atmosphere with global-warming gases. And it will never force massive numbers of people to flee their homes or live in fear of nuclear devastation.

Calling wind power cleaner than other energy technologies is not to say that it is in perfect harmony with the environment. Any energy technology will have some adverse impacts when applied on a scale large enough to provide a significant percentage of America's power needs. Opponents of wind energy tend to highlight those negative impacts without placing them in the broader context of the much more grievous damage caused by our current energy uses.

A vibrant, if at times contentious, dialogue is helping to focus expertise, as well as public attention, on wind energy's environmental costs and benefits. Acknowledging the ways that wind farms can harm the environment is one cornerstone of a successful wind energy enterprise; managing those impacts responsibly is another. Wind developers are now working with a battery of technical experts and civic leaders to get both of those cornerstones in place as they gear up to expand wind's share of America's electricity supply.

CHAPTER SEVEN

Birds and Bats

MANY ASSOCIATE ALTAMONT PASS with the mayhem that ran through a rock music festival at the local speedway in 1969, with 300,000 in attendance and Hells Angels hired as security guards. Unfortunately for wind developers, Altamont has also earned infamy as the place where closely packed clusters of wind turbines have slaughtered tens of thousands of birds, including the federally protected golden eagle.

Over the years, multiple studies have documented the alarming rate of bird mortality at Altamont. In 2004, the Tucson-based Center for Biological Diversity filed suit against Altamont's multiple owners, alleging that 17,000 to 26,000 raptors, or birds of prey, had been killed since the wind energy complex began operating in the 1980s. The primary culprits identified in the suit were the "small, inefficient, obsolete, first-generation wind turbine generators"—with more than 5,000 of them producing only 584 megawatts of rated power.[1] That same year, the California Energy Commission reported the results of a multiyear survey of bird deaths at Altamont. Out of 1,766 to 4,721 birds killed each year, 75 to 116 were golden eagles. The report's authors suggested a number of mitigation actions, but their primary recommendation was to replace the wind farm's low-output turbines with a much smaller fleet of megawatt-plus machines.[2]

In more recent years, the Center for Biological Diversity, five Bay Area Audubon chapters, and other citizen groups have continued to press for tougher bird protection measures, at times working in concert, at times bitterly disagreeing over the adequacy of proposed steps.[3] Finally, in December 2010, the California Attorney General's Office presided over a settlement that committed the wind farm's largest stakeholder, NextEra Energy Resources, to replacing 2,400 aging Altamont turbines with 80 to 120 new ones of equivalent overall capacity by 2015. This effort is likely to be a boon rather than a burden to NextEra, as most of the targeted turbines are already in very poor shape and may be barely functional by the replacement deadline. An official of a smaller Altamont wind company commented at the time of the settlement: "The wind smiths out there hold the turbines together with duct tape and wire."[4]

The killing of birds remains a real concern at wind farms built in recent years, but local citizens and conservation groups have learned to be much more proactive in monitoring wind farm plans as they are being developed, raising objections to ill-chosen sites and intervening early enough in the game to bring about necessary changes. For some years now, Angelo Capparella, an ornithologist by training, has been closely tracking wind development plans in the heavily farmed counties surrounding Bloomington, Illinois, where he teaches systemic and evolutionary biology at Illinois State University. In 2006, he reviewed the draft environmental impact statement for the 100-turbine White Oak Wind Energy Center, a dozen miles northwest of the city—all part of the due diligence on emerging wind projects that he performs for the local Audubon Society chapter. Studying the document, he noticed some glaring gaps. First, he claimed, it failed to give proper attention to adjacent nature reserves. Second, he maintained that it underestimated the number of raptors, including bald eagles, that migrate through the Mackinaw River valley, where the project was to be built. He was further disturbed that no one from the Chicago-based wind development company, Invenergy, had contacted him or, to his knowledge, anyone else at Audubon, Illinois State University, or the Mackinaw Ecosystem Partnership to gather data on local wildlife issues.[5]

Appearing before a county zoning board, Capparella flagged the study's shortcomings, urging the county officials to demand that the wind developer prepare a more thorough analysis. He also invited Invenergy's project manager to his office at Illinois State and showed him the site-specific databases that should have been included in the study. "Your consultant is just looking at easy ways to find data," Capparella chided. He went on to warn the developer that Audubon would not be able to support the wind farm unless significant changes were made. Invenergy immediately responded, adjusting the project's footprint to include a one-mile setback from the Mackinaw River and a half-mile setback from a reservoir surrounded by protected land that provides valuable bird habitat.

Though the outcome of his intervention was positive, Capparella remains wary. Wind projects are moving forward so quickly that there often isn't time to gather reliable data that can help developers really understand how their projects may affect surrounding wildlife. "It is clear, to me at least, the farms are going up faster than the science," he told me as we sat in a motel lounge on the outskirts of Bloomington. I couldn't help feeling that the motel's bland décor matched the monotony of the surrounding countryside, utterly transformed by decades of industrialized farming.

Capparella corroborated my impression that little native biodiversity is left in this vast stretch of central Illinois farm country. Born and raised in North Carolina, he has an outsider's candor when he talks about what's left of the open prairie. He chuckles as he recalls the outrage expressed by some local citizens at zoning hearings on proposed wind farms. "You're ruining my view of the pristine prairie!" they exclaim. His view is decidedly less romantic. "This is all corn and soybean fields; this is *not* pristine prairie! Perhaps people growing up here think this is natural, but it's really quite *un*natural. . . . I just want to protect the remnants that are left."[6]

Changes in turbine technology and wind farm design have substantially reduced raptor kills at modern wind farms like White Oak. At Altamont, the turbines that have posed such a hazard to raptors stand low, with blades circling down to within a few dozen feet of the ground. California's early wind developers packed lots of these machines into limited space, thinking they would be able to harvest more energy from the lands they leased. The combined result was too little room for raptors to maneuver as they pursued rodents and other prey on the ground below. At more modern wind farms, developers have learned to space their machines much more generously to minimize interference with the air currents reaching each turbine. As a rule of thumb, the National Renewable Energy Laboratory (NREL) has recommended that the distance between turbines should be five to ten times the rotor's diameter.[7] That can translate into gaps between turbines of a third of a mile or more for many of today's larger machines.

While this broader spacing is driven by the quest for higher power output, a lucky by-product is the greater freedom of flight that the new design gives raptors and other birds within a wind farm area. The higher ground clearance of larger turbines yields a further margin of safety to raptors swooping down to hunt rodents; the lower reach of blades on today's turbines is well over 100 feet off the ground. And finally, the larger size and slower rotation of modern turbines make them easier for birds to spot and avoid colliding with. Instead of spinning at 30 to 40 revolutions per minute like the older machines at Altamont, large commercial turbines today rotate at about half that speed.

Taller turbines with their longer blades are not an unqualified boon to birdlife, however. Biologist Paul Kerlinger has used radar to track bird migration patterns for more than a quarter-century. He reports that most migrating birds travel between 300 and 2,000 feet above the ground. With the blade tips of today's turbines extending 400 feet into the air and higher, migrating birds in certain areas may be increasingly at risk, especially when flying at night or in low-visibility daytime conditions.[8] Some experts also believe that the pulsing safety lights required by the Federal Aviation Administration may attract migrating birds, particularly on nights with fog or low cloud cover.[9] "These birds orient by the stars, and we hypothesize that they must think the lights are some remnant stars on overcast nights," says Angelo Capparella. Birds tend to fly in circles around these lights in an apparent attempt to orient themselves, he adds. This hugely increases their risk of crashing into swishing blades.[10]

One bird whose fate is being watched particularly closely is the nearly extinct whooping crane. Only a few hundred of these birds remain in the wild, and their migratory path cuts right through the heart of U.S. wind farm country—the Dakotas, Colorado, Nebraska, Kansas, Oklahoma, and Texas. While whooping cranes generally migrate at 1,000 to 6,000 feet, they fly much closer to the ground when looking for roosting or foraging areas. To avert possible harm, ecologist Karl Kosciuch has recommended that 10-mile buffers be created around wetlands that are known to be critical stopover points.[11] No whooping crane deaths have been linked to wind farms, and scientists like Kosciuch are committed to maintaining this perfect record. On the other hand, forty-four whooping cranes have died after colliding with power lines since 1956.[12] As new transmission lines are built to serve an expanded wind power network in the coming years, this hazard may increase.

To ward off bird collisions with turbines, early warning and emergency shutdown systems have been installed at a few wind farms in the United States and abroad. Spanish wind developer Iberdrola uses radar to detect incoming flocks of migratory birds at its giant 404-megawatt Peñascal wind farm in southern Texas.[13] A similar system is being used by AES Geo Energy, a subsidiary of the Virginia-based power conglomerate AES Corporation, at its St. Nikola Kavarna wind farm on Bulgaria's Black Sea coast. Because St. Nikola is sited along a major migratory flyway, the wind farm's environmental management plan requires an ornithologist to be on-site during migration periods, monitoring radar, overseeing fixed-point observers, and coordinating with a mobile tracker to determine when the wind farm's turbines might be placing birds at risk. To make sure wildlife protection is given priority, the decision to shut down turbines rests with the ornithologist, not the wind farm operator. In the fall of 2010, this system proved its worth when strong westerly winds blew more than 30,000 white storks off the sea and toward the wind farm. To protect the flock, ornithologist Pavel Zehtindjiev ordered the turbines to be halted thirty-seven times over a two-day period. Within just five seconds of the "stop" command, the turbines came to a standstill, and the birds passed safely through the wind farm area. Not a single stork was harmed.[14]

Even with improved technology and siting, bird deaths remain a reality at wind farms in the United States and around the globe. Estimates vary, but overall mortality at U.S. wind farms outside California seems to be around two birds per megawatt per year.[15] At that rate, we might end up killing well over half a million birds per year as we approach the Department of Energy's target of installing 300,000 megawatts of wind capacity by 2030—the amount needed to generate about a fifth of our nation's power. While most of the victims are likely to be common songbirds like horned larks, vesper sparrows, and bobolinks, that's still a disturbing number.[16]

Bald and golden eagles, long the focus of controversy at Altamont, have recently become flash points in the dynamics between government regulators and wind developers at the national level. Acting under the Bald and Golden Eagle Protection Act's obligation to prevent the unauthorized "taking" of these iconic birds, the U.S. Fish and Wildlife Service issued a guidance document in February 2011 that outlined a series of steps that wind developers should follow to avoid harming eagle populations. Wind developers vehemently objected to these provisions, claiming that they present enormously cumbersome hurdles that will grind wind farm planning to a slow crawl across the West and, to a lesser degree, elsewhere in the nation. As of this writing, the debate continues.[17]

Wind industry leaders insist that we look at birds killed by turbines within the context of the much more massive toll on bird life that we seem to accept quite readily, all in the name of comfort, convenience, and affinity for felines. I think back to my meeting in Denmark with Vestas communications chief Peter Kruse. "Your neighbor's cat is a far more efficient killing machine than my turbines," he bantered as we surveyed the issues that wind energy opponents are prone to raise.

Behind this flip remark lies a body of research that, over decades, has tried to gauge the risks to bird life from a wide range of causes. Every year in America alone, collisions with buildings may kill anywhere from 97.6 to 976 million birds, and road strikes may cause an additional 80 million deaths. Up to 50 million birds die when they crash into communication towers, and 130 to 174 million birds are killed when they collide with high-tension, or "bulk," transmission lines.[18] Beyond those hazards, there are many other ways we end up killing birds and destroying their habitats through energy industries like oil and gas drilling and refining, mountaintop removal and strip mining for coal, and power plant operations. Farming wreaks further havoc on bird populations by radically altering natural landscapes and saturating field crops with pesticides. And then, of course, there are Kruse's cats. The Fish and Wildlife Service estimates that hundreds of millions of songbirds and other bird species are hunted down and killed by domestic and feral cats in the United States every year.[19] Researchers at the University of Georgia have arrived at an even higher number: they say that "free-ranging domestic cats," themselves numbering between 117 and 157 million, kill more than 1 billion birds annually in the United States.[20]

Killing birds is one concern; disrupting their habitats is another. In Wyoming, wind energy proponents have found themselves at the center of a superheated controversy over the greater sage grouse, a large, richly feathered fowl that has long hovered on the brink of being listed as an endangered species by the U.S. Interior Department. A century ago, the sage grouse numbered about 16 million, its habitat spreading across sagebrush country throughout most of the American West, extending north into Alberta and Saskatchewan. Today, its population has dwindled to a few hundred thousand, with much of its remaining habitat in large sections of Wyoming where oil, gas, and coal development have already encroached heavily.[21]

By the time wind developers began looking seriously at major new wind farm sites in Wyoming, conservation groups in the West had already launched efforts to get the sage grouse listed under the Endangered Species Act. The Interior Department ruled against listing the species in 2005, arguing that the bird's population had increased or stabilized in several states since 1985. This led to a federal lawsuit that yielded a stern court ruling in 2007, criticizing the lack of scientific rigor underlying the department's decision and requiring it to reconsider the matter.[22]

Wyoming politicians and energy industry leaders then grew alarmed. Recalling the federal listing of the northern spotted owl as a threatened species in 1990, they knew how profoundly that decision had affected logging in the Pacific Northwest. A federal listing of the sage grouse, they feared, would shut down much of Wyoming's extremely lucrative coal, gas, and oil industries. To avert a crisis, Governor Dave Freudenthal convened a multi-stakeholder Sage Grouse Summit in June 2007, where he stressed the urgent need for a state conservation plan strong enough to fend off the bird's federal listing.

The following year, Freudenthal signed an executive order creating a Greater Sage-Grouse Core Area Protection plan. This measure gave protected status to about a fifth of Wyoming's territory, thought to contain more than 82 percent of the state's sage grouse population. Within the designated "core areas," existing energy facilities—oil and gas wells, pipelines, and even coal mines—were allowed to continue operating. New projects, however, bore the burden of proving that they would not diminish the sage grouse population.[23]

Wind farms were particular targets of this requirement. Unfortunately, they also became a focus for the governor's open disdain. He accused wind developers of being driven by a "gold rush" mentality that placed "[s]eemingly every acre . . . up for grabs in the interest of 'green, carbon-neutral technologies' no matter how truly 'brown' the effects are on the land." He also likened wind developers' zeal in siting new projects to "taking a shortcut to work through a playground of school children and claiming 'green' as a defense because you were driving a Toyota Prius."[24]

In a state whose landscape has been punctured by oil and gas wells, scarred by pipelines, and torn apart by the nation's largest open-pit coal mines, Freudenthal's anti-wind rhetoric was surreal. But with more than half of the state's revenues coming from energy-related taxes and royalties,[25] it's hardly surprising that Wyoming's chief executive would go to great lengths to protect that flow of income. In a second order issued in 2010 and now in effect, the bias toward the state's traditional energy industries is even more unmistakable. Not only are existing energy operations with their drill sites, pumping stations, access roads, and vehicle traffic allowed to continue, but new oil and gas wells are permitted in core sage grouse areas at a density of one drill pad per square mile. Although it's hard to fathom why, mines are also allowed at the same average density. Wind development, by contrast, is "not recommended."[26]

The actual impact of wind farms on the sage grouse remains unclear. Experts suspect that the birds may stay away from wind turbines, fearing that raptors—their natural predators—are using them as perches. They also worry that power lines, guy wires on meteorological towers, and turning turbine blades may pose collision risks.[27] Few studies have been conducted to probe these and other issues, lending credence to Angelo Capparella's observation that wind farm development is several steps ahead of the available science.

Wyoming's brinkmanship, going tough on wind but easy on mineral extraction, is destined to become harder to sustain as the sage grouse moves ever closer to earning federal endangered species status. In March 2010, Barack Obama's Interior secretary, Ken Salazar, distanced his agency from the George W. Bush administration's decision not to consider the bird endangered. On one hand, he announced that listing the sage grouse was "warranted but precluded" in light of other species more urgently needing federal protection. On the other hand, he designated the bird a "candidate species" whose status is to be reviewed annually, clearly signaling close ongoing scrutiny and the very real possibility of a future listing.[28]

Erik Molvar, a wildlife biologist who directs the nonprofit Biodiversity Conservation Alliance in Laramie, objects strongly to wind's uneven treatment under Governor Freudenthal's sage grouse protection program. "There's nothing fair about it," he says, pointing to the oil and gas industry's enormous influence on state and federal politics. "The oil and gas industry should be held to the same bar as the wind energy industry," he says, "but it should be that higher bar of being compatible with protecting public lands and wildlife."

Fortunately for Wyoming wind developers, possible sites for new wind farms abound outside the state's designated core sage grouse areas. Molvar makes this point as he leads me through *Doing It Smart from the Start*, his organization's guide to responsible wind energy development. The report overlays maps of the state that show critical ranges and migration corridors for birds, bats, and multiple big-game species including antelope, elk, bighorn sheep, and mule deer. It also looks at historic trails and other visually sensitive areas. Putting it all together, the report arrives at a composite map showing where wind power should and should not be developed. The results, as Molvar describes them, are encouraging. "There are 5 million acres of 'go-zones' in Wyoming, and of those 5 million acres, 4 million acres have high wind potential." That's far more land than the state would need to achieve its maximum build-out of wind turbines, he assures me.[29]

The greater prairie chicken is another member of the grouse family that has entered the debate about wind farm development. Once there were three subspecies of this bird. One is now extinct; another survives only in small areas of southeastern Texas; and the third continues to populate sections of mixed and tallgrass prairie in Kansas and several other Midwestern states.[30] While greater prairie chickens have decreased dramatically in overall numbers as cultivated farmlands have encroached on the tallgrass prairie, they are not listed as endangered and continue to be legally hunted. Environmental biologist Robert Robel refuses to give an estimate of how many of this species remain in Kansas, but he calls the population "healthy" and insists that hunting prairie chickens—a popular pastime—does not pose a threat to their survival. He points to intrusions that interfere with mating and nesting as having a much greater impact on the bird's ability to maintain stable numbers. That's where wind farms come into play.[31]

Male prairie chickens compete with one another for sexual dominance through a mating dance called "booming," which involves lots of shuffling, strutting, neck-puffing, and cooing. They carry out this ritual on open ground where there are no nearby trees or other high structures that could host birds of prey. Like the scientists who are studying the sage grouse, Robel wonders whether wind turbines cause prairie chickens to abandon nearby mating grounds. When I spoke with the retired Kansas State University professor in January 2010, he was midway through supervising a study of greater prairie chicken activity at the Meridian Way Wind Farm, whose footprint includes about 13,000 acres of mixed-grass ranchland along with several thousand acres of cultivated crops. While no results about the effects of Meridian Way's turbines were available at the time, Robel pointed to his own earlier collaboration on a study that looked at how other built structures affected a somewhat smaller relative, the lesser prairie chicken. Researchers in that study found that nesting birds stayed away from buildings, gas compressor stations, and a coal-fired power plant, and they further observed that nesting was unlikely to happen within 1,300 feet of transmission lines or paved roads.[32]

When Horizon Wind Energy settled on Cloud County as the site for Meridian Way, its environmental team recognized that the wind farm might harm local grassland habitats. The company, therefore, took the unusual step of committing to buy up conservation easements on an expanse of grassland equivalent in scale. It turned to the Ranchland Trust of Kansas, an offshoot of the Kansas Livestock Association, as its primary partner in negotiating easements with landowners in the Smoky Hills, a few dozen miles south of Meridian Way. With an undisclosed sum of money at her disposal, the trust's Smoky Hills coordinator, Stephanie Manes, began having conversations with area landowners. In signing a conservation easement, landowners would receive a one-time payment in exchange for maintaining their properties as grassland in perpetuity. They could continue to graze cattle and live on the land, so long as they adhered to stipulated practices for grazing, fencing, and weed control. They could not, however, convert pastures to cropland or sell off any of the property for development. And, of course, they could not put up any wind turbines.

The agreement between Horizon and the Ranchland Trust called for 13,100 acres to be placed under conservation easement, out of a total of 20,000 acres that were to be restored as grassland bird habitat. Through smart leveraging, Manes quickly pushed beyond that original goal. Within nine months, she had lined up 25,000 acres of easements and expected to bring another 5,000 acres under conservation protection with her remaining funds.

Manes is careful not to take a categorical stand on wind farms. She makes it clear that the Ranchland Trust's umbrella organization, the Kansas Livestock Association, has remained neutral on the subject. "Our membership is pretty much split evenly down the middle," she says of Kansas ranchers. She lets me know that she has her concerns, though. From her training in wildlife management and grassland ecology at Oklahoma State, she knows what a powerful deterrent tall trees can be to grassland bird mating and roosting on the open prairie. "One tree per ten acres to a human is like, 'Oh, what a beautiful tree!'" she explains. "Well, that tree will take out the entire ten acres of habitat for grassland nesting birds." Will grassland birds be similarly deterred by wind energy arrays, or will they learn to see turbines as non-threatening? Manes hopes that the study at Meridian Way, supervised by Robel, will begin to answer that question. "They just haven't been around long enough for us to know," she says of the wind turbines.[33]

Though uncertain about wind turbines' impact on grassland birds, Manes has no doubt about the harm caused by another human activity: the annual burning of the Kansas tallgrass prairie by ranchers. She points in particular to the Flint Hills, where land agent Jim Roberts ran into such fierce opposition when he was scouting for possible wind farm sites on behalf of Horizon's predecessor, Zilkha Renewable Energy. Every spring, ranchers wielding kerosene torches race through the Flint Hills on their all-terrain vehicles, setting about 1.7 million acres of grasslands aflame. Thinking of this conflagration in New England terms, I am stunned to realize that it covers an area more than twice the size of Rhode Island.

To some degree, periodic burning is essential to the prairie's survival, especially in regions like central and eastern Kansas, which experience more rainfall than the semiarid sagebrush prairie to the west. Without fires to wipe out seedlings, larger plants, including trees, would invade quickly, converting prairie to forest in a matter of years. Long before white settlers arrived, Plains Indians relied on fires—some natural, some set—to maintain grasslands for their bison herds. Scientists today estimate that grassland fires are needed every three to four years to keep the prairie intact.[34]

Today's Flint Hills ranchers have pushed far beyond this threshold. Trucking in young cattle from out of state for a few months of intensive grazing every spring, they know that their heifers will gain weight much more quickly when they graze on the fresh grass that comes up after a burn. Quick weight gain translates into bigger profits, especially for those who double-stock their ranchlands, running two cohorts of cattle through shorter cycles in a single grazing season.

It's no small irony that Flint Hills ranchers, adamantly opposed to wind turbines in their midst, resort so readily to annual burning—a practice that Stephanie Manes calls an ecological nightmare. They don't seem troubled that their blazes create enough air pollution to trigger ozone alerts in Wichita, Topeka, and other urban areas.[35] They also seem unbothered that the fires rob prairie chickens and other grassland birds of the thatch they need to build their nests and the ground cover they rely on to hide their young from birds of prey. John Briggs, who directs the Konza Prairie Biological Station just outside Manhattan, Kansas, quantifies how much camouflage is needed to protect nesting birds: "If you throw a football into a field and it lands and you can't see it, that's enough cover for prairie chickens," he tells me.[36] After the spring burn, a billiard ball wouldn't pass this test.

Professor Robel and colleagues from Kansas State have looked at the effects of springtime burning on grassland bird nesting in eastern Kansas. Their results were stark, if predictable. On unburned fields, they found 372 nests. On burned fields, they spotted only 27 of them.[37]

The verdict is still out on whether, and to what degree, Meridian Way's wind turbines have disrupted prairie chicken habitat in Cloud County. Whatever the impacts turn out to be, they will surely pale when compared to the devastation wrought by other human intrusions. Horizon's chief environmental officer, Rene Braud, makes no attempt to disguise her frustration with the degree to which wind developers are being called to account for environmental harms to ecosystems long ravaged by other human activities. She speaks of the Kansas prairie in much the same way Wyoming wind advocates describe their own state's eroded sagebrush terrain: "There's no question it's been decimated by agriculture, cattle, oil and gas, housing, and other forms of development," she says, noting how unfair she feels it is to single out the wind energy industry for heightened scrutiny. "We're taking the hit for the last two centuries."[38]

From the early days of wind energy development, it was obvious that birds were being killed by turbines. The toll on bats was initially less apparent. Even today, experts are careful to point out that most of the data on bat mortality at wind farms has come from surveys that were primarily focused on birds.[39]

Available studies may still be limited, but they reveal a level of bat mortality that is of real and mounting concern. Studies conducted at wind farms in the West and Midwest point to annual bat kill rates ranging from 0.8 to 8.6 per megawatt—more or less in line with reported death rates for birds. Along some forested ridgelines in the East, however, the losses have been much higher, reaching as high as 41 bats per megawatt at the Tennessee Valley Authority's Buffalo Mountain wind farm near Oak Ridge, Tennessee.[40]

What draws bats to turbines is uncertain. Some experts assume that they come to feed on insects congregating in the clearings surrounding turbines; a related hypothesis is that the insects themselves are lured by the heat given off by turbine machinery. The mechanical or aerodynamic sounds that turbines produce may also appeal to bats. And then there's the possibility that bats confuse turbines with tall trees suitable for roosting.[41]

Whatever the attraction, bats are killed by wind turbines in two primary ways, as evidenced by carcass surveys. Some collide with rotors; apparently the radar that normally guides bats' flight has trouble detecting blades in motion. Most deaths don't result from collisions, however. Roughly 90 percent of bats are killed by internal hemorrhaging, or barotrauma, caused by the rapid changes in atmospheric pressure that occur within the vortices of spinning blades.[42]

While no recorded deaths of threatened or endangered bat species have been linked to turbine operations,[43] the development of one wind energy project—the Beech Ridge Wind Farm in Greenbrier County, West Virginia—has been curtailed to prevent possible harm to the Indiana bat, listed as endangered since 1967. The Indiana bat is tiny, weighing about a quarter of an ounce and with a body 1.5 to 2 inches long. Its habitat extends across much of the East and Midwest, though its numbers have dropped by more than 50 percent since it was first listed as endangered. During the warmer months, Indiana bats live in wooded and semiwooded areas, roosting under tree bark and in dead trees. In winter, they hibernate in caves.

Beech Ridge Energy LLC, a subsidiary of Invenergy, announced plans to build its wind farm along a 23-mile stretch of Appalachian ridgeline in November 2005. As part of the project preparations, a consultant examined caves and conducted mist-net surveys to check for the possible presence of Indiana bats in the area. The consultant's reports showed three caves currently used by hibernating Indiana bats between 5 and 10 miles from the project site, but none within 5 miles of the site. No Indiana bats were captured by the nylon-mesh mist nets, strung up like badminton nets at multiple locations. Based on these findings, Beech Ridge sought and received permission from the West Virginia Public Service Commission to build its wind farm: 124 turbines totaling 186 megawatts of installed capacity.

Unhappy with the state agency's decision, local citizens joined forces with a national nonprofit, the Animal Welfare Institute, in a suit claiming that already-dwindling colonies of the Indiana bat would be further diminished by the wind farm in violation of the Endangered Species Act. Federal judge Roger W. Titus agreed, invoking Ben Franklin when he ruled that "like death and taxes, there is a virtual certainty that Indiana bats will be harmed, wounded, or killed imminently" by the wind farm.[44] Looking closely at the work performed by Beech Ridge's consultant and hearing testimony from other bat experts, he chastised the consultant for failing to consider acoustic monitoring data that indicated the likely presence of Indiana bats in the project area, and he took issue with the consultant's assumption that bats hibernating in caves 5 or more miles from the site would be unlikely to approach the wind farm. Other experts testifying in the case maintained that migrating bats flying to or from caves some distance from the wind farm could easily find themselves crossing through turbine zones.

The judge's order, issued in December 2009, barred Beech Ridge Energy from adding to the forty turbines already built and confined the wind farm's operations to the winter months, when the bats would be in hibernation.[45] In January 2010, the parties reached a settlement agreement—authorized by the court—that expanded the permissible number of turbines to sixty-seven. The agreement also modified the winter-only restriction, allowing the wind farm to operate during daylight hours at other times of the year.[46] Meanwhile, Beech Ridge Energy has applied to the U.S. Fish and Wildlife Service for an "incidental take" permit under the Endangered Species Act. This permit, if granted, might allow Beech Ridge to operate during nighttime hours so long as it complies with the terms of a habitat conservation program designed to minimize "incidental" deaths and other injuries caused to the Indiana bat.

Apprehensions may be greatest regarding endangered species like the Indiana bat, but wind companies are generally concerned about the toll that turbines are taking on bats, and they are experimenting with ways to reduce that toll. Some are weighing whether to take turbines out of service during the two hours after sunset, when insects are most abundant and bats are most likely to be pursuing them. Others are testing whether bat deaths can be reduced by shutting down turbines during low-wind periods, when bats and insects are much more likely to be flying than in higher-wind conditions.

Don Furman at Iberdrola Renewables introduced me to his company's collaboration with a nonprofit group, Bat Conservation International, at the Casselman Wind Power Project in western Pennsylvania. During periods of peak bat activity over a two-year period, Iberdrola operated its turbines at three different "cut-in speeds"—the minimum wind velocities at which rotors are allowed to start turning. At the low end was the turbines' normal cut-in speed: 3.5 meters per second, or 7.8 miles per hour. In the mid-range, turbines were stalled until the wind speed reached 5 meters per second (11.2 miles per hour). At the upper end, the turbines were allowed to operate only if the wind was blowing at a minimum of 6.5 meters per second (14.6 miles per hour). The research team found that, by raising the cut-in speed, bat fatalities could be reduced by 44 to 93 percent.[47] Iberdrola also found that these curtailments, implemented during the peak bat migration period from late summer to early fall, caused relatively minor losses in power production: 0.3 percent of total annual output when the cut-in speed was 5 meters per second, and 1 percent when it was 6.5 meters per second.[48]

It is hard to arrive at a reliable forecast for overall bat deaths as we build more wind farms in the years ahead. Technology may change; wind farm operations may become more responsive to surrounding conditions; and wind farm siting may increasingly avoid bat population hot spots. Projecting from current mortality rates and assuming no technology or design improvements, one study estimates that the annual death toll in the Mid-Atlantic Highlands alone could range from 33,000 to 111,000 bats by 2020. Nationwide, the total would likely be much higher, although bats generally have been found to be less vulnerable to wind turbines sited in unforested areas of the West and Midwest.[49]

There is no denying that wind energy takes a toll on birds and bats, but there is also no question that the continued burning of coal and other fossil fuels exacts a much greater environmental cost. Beyond the localized devastation that accompanies our extraction and use of fossil fuels, we need to look at the utter transformation of our global environment—the destruction of entire ecosystems, the wholesale elimination of species, the reduced availability of freshwater resources, and the massive displacement of human populations—that will result from our failure to tame the global warming juggernaut. Wind energy technologists certainly should commit themselves to coming up with better ways to detect and respond to the presence of birds and bats at wind farm sites, just as government regulators and concerned citizens should demand rigorous planning and vigilant operation of wind farms (as well as all other energy facilities) to minimize damage to wildlife. The U.S. Fish and Wildlife Service's eagle conservation guidelines may require further streamlining to ensure necessary protections in a manner that doesn't stymie responsible wind development, but its leadership in this area bodes well for a more consistent approach to wildlife protection at wind farm sites.[50]

Managing wind energy's wildlife impacts is essential. If we fail to address them in earnest, we risk derailing a technology that, wisely implemented, can set us on a new energy course that will vastly reduce our local and global environment's exposure to much more fundamental dangers.

CHAPTER EIGHT

The Neighbors

I ONCE TOLD A NEWSPAPER REPORTER that I find wind turbines beautiful. She responded sharply that, to her, they're ugly. Other than the noise turbines make, nothing about wind farms creates greater public discord than their visual appearance. Embroiled in the battle over building an offshore wind farm in Nantucket Sound, Cape Wind developer Jim Gordon often reassured skeptics that the proposed turbines would appear no higher than a thumbnail held at full arm's length when viewed from the nearest landfall. Anti–Cape Wind activists countered by decrying the industrialization of their ocean horizon.

In the 1980s, most Americans would have associated wind farms with the chaotic turbine arrays that crowded California's Altamont, San Gorgonio, and Tehachapi passes. In their rush to build, developers purchased a haphazard variety of turbines, some with two-bladed rotors, others with three, some mounted on spindly steel-truss towers, others capping smooth metal tubes. The visual effect was jarring and, to some, alien. "Spielberg and Lucas could not have done better," urban and regional planner Sylvia White wrote to the *Los Angeles Times*. "Once-friendly pastoral scenes now bristle with iron forests," she fumed.[1] Veteran wind advocate Paul Gipe hyperbolically likened the altered landscape at one California site, with its crude access roads gouged into mile after mile of barren hillsides, to parts of Appalachia where coal mining via mountaintop removal has devastated the natural topography.[2]

Today's wind farms are generally much kinder to the eye and more respectful of the landscape. Unlike the early California arrays, almost all modern wind farms stick to a uniform turbine profile, even where more than one manufacturer's equipment is used. Wind farm layout also tends to be much more orderly, with turbines generously spaced and careful consideration given to the turbines' appearance as part of a broader visual experience. Even with these changes, the much larger size of today's turbines undeniably makes them visually dominant at close range and noticeable from afar in many settings. One of those settings is Nantucket Sound, where my home state's long-proclaimed renewable energy ambitions ran headlong into the aesthetic passions of well-heeled vacationers, including our state's most revered political hero.

I will never forget the sunny morning in April 2006 when a few dozen of us gathered outside Boston's historic Faneuil Hall. Senator Ted Kennedy had come to Boston to witness Governor Mitt Romney's signing of the Massachusetts health-care reform bill, and we were there to greet him. As the senator stepped briskly out of his black SUV, he faced a chorus of chants and a sea of red-white-and-blue posters showing wind turbines rising above three bold letters: YES. He rushed into the building, looking sheepish and stunned.

The Conservation Law Foundation, which I headed up at the time, usually didn't participate in public demonstrations. We preferred more orderly professional settings where our attorneys could best apply their skills. However, Kennedy's backroom politicking to block the Cape Wind project had pushed us over the edge. We weren't willing to let him undermine this pioneering project.

During the preceding months, Kennedy had recruited Senator Ted Stevens and Representative Don Young, both Alaska Republicans, to help him blow Cape Wind out of the water. Stevens served on a House-Senate conference committee charged with finalizing the Coast Guard's annual reauthorization bill; Young was the committee's chair. A veil of secrecy surrounded the committee's deliberations, but word filtered out about the machinations that were taking place behind closed doors. First, Representative Young introduced language barring offshore wind turbines within a mile-and-a-half of any navigational channel—a much bigger safety buffer than the federal government had ever required for oil rigs and other ocean installations.

The measure's unstated goal was to reduce the scale of the Cape Wind project to such an extent that it would no longer be financially viable. When the media exposed this ploy, public furor mounted and several legislators objected that the measure violated basic principles of good governance. Then Senator Stevens stepped in, substituting new language giving governors of adjacent states veto power over the wind farm. Even though all of Cape Wind's 130 turbines were to be sited in federal ocean waters outside any state's jurisdiction, this provision would have allowed Governor Romney—an avowed opponent of the project—to stop the development in its tracks.

I personally was deeply upset with Senator Kennedy for orchestrating these maneuvers; they seemed woefully out of character for a politician who had, for so many years, been a steadfast champion of policies that would open up the U.S. marketplace to renewable energy. It felt like a self-serving betrayal, all to preserve an unobstructed ocean vista from the Kennedy vacation compound in Hyannis Port, more than five miles from the nearest proposed wind turbine.[3]

Wanting to get a better idea of how a project like Cape Wind might look to Massachusetts vacationers, I made it a priority to visit one of the world's largest offshore wind facilities when I traveled to Denmark in October 2009. The Rødsand Offshore Wind Farm lies just off Denmark's southern coast, in the Baltic Sea. I set my sights on the village of Nysted, whose snug harbor is a major attraction to Baltic boaters. Parking near a row of well-maintained brick houses, I walked to the water's edge, where a few dozen sailboats and motor cruisers were tied up along spindly wooden piers, their prows packed together like cars in a busy parking lot. I strained to look beyond the harbor, but I found myself distracted by the motor cruisers' flying bridges and the sailboats' tall masts with their wire stays twitching in the breeze. In the distance, about six miles from the shore, I could vaguely pick out a hazy sequence of finely drawn vertical shapes in evenly spaced rows, barely rising above the horizon. So these were the industrial intruders that Ted Kennedy found so offensive?

When I visited the Rødsand wind farm, its first phase had been in operation for six years, with 162 turbines capable of delivering 373 megawatts of power. The second phase was under construction. A little over half as large as the first phase, Rødsand II was expected to meet the power needs of up to 200,000 Danish homes. Bjarne Haxgart, project manager for E.ON Climate and Renewables, described some of the challenges involved in building the project. To protect porpoises from the harsh sounds made by pile-driving, his crew was using an acoustic alarm to drive them away from the construction zone. This evacuation was only temporary, though. Porpoises have been observed returning to offshore wind farm construction areas within hours after pile-driving operations cease,[4] and Haxgart says they are commonly spotted in the waters surrounding Rødsand's turbines. Fish also flourish around the wind farm, Haxgart notes, with the turbines' large concrete foundations serving as artificial reefs.

As with land-based wind farms, offshore projects like Rødsand may create some level of disruption to bird life in the area. Long-tailed ducks—a particular focus of concern among conservationists—may move away from the immediate vicinity of the wind farm, but the environmental impact study commissioned by E.ON predicts few bird collisions with turbines and no significant impact on the international population of this species.[5]

Turning to the wind farm's visual impacts, Haxgart took me to a large plate-glass window at the Rødsand field office, just across a short stretch of salt marsh from the shore. He handed me a pair of binoculars and pointed to where crane-equipped barges were lowering premolded concrete turbine foundations into the sea. Even with this visual aid, the turbine emplacements were so distant that I found it hard to make out what was happening. Haxgart then explained that the wind farm's visual impacts are greatest at night, when the aviation safety lights are flashing. Initially the turbines were equipped with red lights, synchronized via satellite to trigger simultaneously. On occasions when they went out of synch, people would get upset. "It looked like Las Vegas at nighttime," he acknowledged.

With permission from the Danish aviation authorities, E.ON has replaced the blinking red beacons with a white strobe and a fixed, low-intensity red light. Apparently this is less jarring to the wind farm's neighbors.[6] Even so, the environmental impact study for Rødsand II describes the wind farm's expected nighttime appearance as "comparable to the effects of a small town," particularly from the vantage point of its closest landfall, just over a mile from the nearest turbine.[7]

Hearing about the lights at Rødsand brought me back to my first glimpse of the Grand Ridge Wind Farm in Illinois. It was a clear evening in early May as I approached Marseilles on Interstate 80, heading west from Chicago. A few miles before the exit, I looked to my left and there, hovering in the distant darkness, was an eerie constellation of red lights, all blinking in unison. I later checked the map: Grand Ridge's closest turbines were ten miles from the highway.

The next morning I drove through Marseilles, a small town with a few fading industries on the banks of the Illinois River, and climbed up a short wooded rise onto the ridge that gave the wind farm its name. Seeing the turbines in the bright morning light came as a relief. They were generously spaced across the gently undulating fields, punctuating rather than crowding the horizon. Their profiles blended gracefully with the farmhouses and barns, some of peeling clapboard, some with aluminum siding, almost all of them white like the turbines. The turbines' clean geometric lines handsomely offset the ever-present silos, perfectly cylindrical, their galvanized steel glinting in the sun. And they were far less unsightly than the rusted skeletons of water-pumping windmills that still stand in so many barnyards.

The Grand Ridge Wind Farm wasn't the first power producer to reach this stretch of central Illinois farm country. Just a few hundred yards from the nearest turbine was the LaSalle County Nuclear Station, its boxy, windowless brick shell and candy-striped smokestack jarring in their juxtaposition to the adjacent fields of corn stubble. The sluiceway that carried cooling water in a long steaming arc from the plant to a large artificial pond also struck me as an uneasy contrast with the surrounding farmlands. A sign welcoming visitors to the LaSalle Lake State Fish and Wildlife Area did little to conceal the true function of the pond, with its ruler-straight borders and power plant looming in the background.

I met with several farm families living in the midst of Grand Ridge's sixty-six turbines and heard very different attitudes about the wind farm. Frank and Sarah Diss told me that the two turbines on their 169-acre farm are easy neighbors, neither noisy nor visually intrusive. Beyond that, they're thrilled to be part of the green power revolution. "We got on the bandwagon a little bit," Frank told me. "Wind power, green energy—we have to have it." Pointing at a wind tower on the far side of his barn, about a quarter-mile away, he said: "One of those will provide enough electricity for 400 homes, so we don't have to fight foreign oil."

Bob and Ruth Widman, who live a mile or two away from the Disses, are much less happy having turbines in their midst. They have lived in the same house since 1955, and their son continues to farm the property. "It changes the whole rural area," Ruth delicately commented. The turbines distract and disorient her when she's driving, and she welcomes the visual relief when she makes the occasional trip away from her home turf. "It seems so good to get out someplace where we can look and see farms," she sighed. Bob spoke about the turbines in cruder and angrier terms: "They're like cow manure—they're all over the place."[8]

In my visits with the Disses, the Widmans, and others, I found it hard to separate their divergent perspectives from their very different financial stakes in the project. The Disses get $8,250 in annual lease payments for each of the two turbines on their property, plus a dollar per year for each linear foot of access roads crossing their land.[9] I assume that their enthusiasm about wind energy is reinforced, at least to some degree, by those benefits. The Widmans initially considered joining the project, but Bob bargained for higher lease payments than the project's developer, Invenergy, was willing to offer. He told the company's land agent that he wasn't willing to take less than $10,000 per turbine per year, so they ended up outside the project, earning nothing while looking out at their neighbors' turbines.[10] I'm not surprised that they now feel alienated.

At Grand Ridge, as at dozens of other locations across America, wind farms may not be universally embraced, but they are fast becoming part of the working landscape. They are practical implements, akin to tractors and harvesters, even if their physical presence is more dominant and dynamic. Where wind developers run into stiffer opposition is in areas where their projects are seen as diminishing highly valued natural or historic landscapes. The Flint Hills of Kansas are certainly one such place.

Rose Bacon grew up on a farm in Iowa. When she was about twelve, her father brought her along on a trip to the Flint Hills to buy cattle. It was love at first sight, she recalls. "I told my dad on the way home, 'That's where I belong.'"[11] It took a few decades to realize her dream, but in 1991, she and her husband, Kent, bought a 520-acre ranch in the heart of the Flint Hills, just south of the village of Council Grove. Since then, they have grazed up to a thousand head of cattle on their rich grassland pastures each summer, trucking them back to their out-of-state owners or sending them on to a feedlot after they've had their fill of prairie grass. Kent, a Vietnam vet, has a specially contoured prosthetic leg that allows him to sit comfortably in the saddle as he works the herds on horseback.

It may be Rose's outsider perspective that has made her such a passionate protector of the Flint Hills. "When you move to a new area, you perhaps don't take it for granted the way you might if you grew up in that particular place," she observes. Iowa once was prairie; now it is cultivated farmland. She knows how utterly a landscape can be changed by human industry, and in her view, wind energy would bring just that sort of unwelcome transformation to the Flint Hills. Once the tallgrass prairie extended across 170 million acres of the eastern Great Plains. About 4 percent remains today, with two-thirds of that in the Flint Hills, extending in a 50-mile-wide swath from just south of the Nebraska border all the way through Kansas and down into Oklahoma.

I visited RK Cattle, Rose and Kent's ranch, in the midst of a late-July heat wave. After serving me a bacon-and-eggs breakfast, Rose asked if I would be more comfortable touring the ranch by horse or on an all-terrain vehicle. I opted for the horse but admitted to being a bit nervous; I had last been on horseback as a camper in the 1960s. After a brief riding lesson, we headed up into the hills. Rose pointed to the rich layer of grasses and wildflowers beneath our horses' hooves and rattled off names like bluestem, buffalo grass, and ironweed—just a few of the plants that inhabit this landscape. The grasses were still green, but by autumn they would be tall and brown, kept alive by roots that grow several feet into the rocky soil. I gazed all around me. Nothing more prominent than the occasional barbed-wire fence broke the flow of the grassy uplands. The only trees in sight were far below us, running along the path of a narrow creek.

Down by the creek, Rose showed me a small grotto of layered flintstone ("chert" is the more accurate term, she told me) where water trickled into a shallow pool no more than a dozen feet across. I called it a spring; she corrected me, telling me it's a waterfall. "The scale here is not like the Rockies—we don't have thousand-foot waterfalls," she said. Then she quoted a well-rehearsed adage about the Flint Hills: "It doesn't take your breath away, but it gives you a chance to catch your breath."

Rose's voice stayed calm as she shifted to talking about wind farms, but her steely determination as an anti–wind farm fighter came through in her choice of metaphor. "I personally look on the wind complexes as a rape of the landscape, and I don't use that term lightly." She reminded me that, before coming to Kansas, she worked as an emergency-room nurse. "You see those people come in. They are battered and shattered and torn up. There's no doubt in your mind what happened." Then, she continued, a year or more may pass before the trial. By that time, the obvious scars will have healed, though not the deeper physical and psychological wounds. "And the defense will say, 'See? Basically there was no damage here. Probably it was invited.'" She likened this to the scars in the landscape caused by wind farm access roads and concrete-slab turbine foundations. Wagon-wheel ruts from the Santa Fe Trail are 150 years old, but you can still find them. Likewise, she said, the subterranean remnants of turbine footings may prevent prairie tallgrasses from sending roots deep into the soil, stunting their growth. "You're going to have these little rocky, bare, weedy spots where everything dies on days like this." These wounds, she warned, will last long after the turbines have come down and the viewscape has been restored.

Rose Bacon's romance with the Flint Hills landscape unnerved me. Her self-avowed quest to preserve its pure prairie essence left no room for accommodating certain human intrusions, yet she seemed to accept others readily. She likened wind farms to the most heinous of criminal insults, but she dignified the burning of more than a million-and-a-half acres of tallgrass prairie as an "annual rite" that ushers in the renewal of lush green grass and countless varieties of wildflowers.[12] What about the prairie chickens and other grassland birds that the flames rob of their nesting grounds, and what about all that air pollution, sufficient to set off ozone alerts in cities dozens of miles away?

I was also bothered by Rose's apparent need to invalidate just about every possible argument for wind energy. She dismissed Al Gore's "Chicken Little theory" about climate change as "a load of bull" and derided wind technology as inefficient and unreliable. She railed against the subsidies given to wind developers, ignoring our government's long history of providing much more massive support to the fossil fuel and nuclear industries. And she extolled the beautiful pictures she'd seen of reclaimed surface coal mines. Could she really imagine that those gaping canyons, blasted, eviscerated, and bulldozed, were less devastating to the environment than a few dozen buried turbine footings spread across several miles of open landscape?

Ron Klataske, who directs Audubon's Kansas chapter, sees wind energy in less black-and-white terms than Rose Bacon. He acknowledges that some wind farms in Kansas are well sited, pointing in particular to two completed projects in the heavily farmed southwestern corner of the state. He also recognizes that the Flint Hills are a tempting target for wind developers, given the stiff, steady winds that blow through the area and the proximity of the Flint Hills to cities in eastern Kansas and over the border in Missouri. Yet he shares Rose's antipathy to wind in the Flint Hills and has been fighting what his organization has called "a potential tsunami of industrial-scale wind turbine complexes" and "a killing field for migrating birds." He hopes that, over time, the U.S. Fish and Wildlife Service will acquire a million acres of conservation easements in the Flint Hills. In the meantime, he has been encouraging the purchase of conservation easements by groups like the Ranchland Trust of Kansas and the Nature Conservancy.[13]

Ron arranged for me to meet some of his allies in the Flint Hills campaign against wind. We shared beer and burgers at the Hitchin' Post in the village of Matfield Green, on the Flint Hills Scenic Byway. Then we drove east on a dusty gravel road, crossing a stretch of soybean fields before reaching the open prairie. Just as the sun was setting, we pulled off the road on the crest of a windswept hill. Broad stretches of undulating green surrounded us, softened by the fading light. Thinly scattered herds of cattle peppered the landscape—Black Angus, mainly. A nighthawk hovered overhead, crickets raged, and a coyote howled in the distance.

Bill Browning, a physician who still lives on his family's multigenerational ranch, opened the conversation. "The beauty of it, for me, is where the hills meet the sky, morning and evening, and the shadows come across the hills and make all the contours stand out. If you're going to put a string of 400-foot steel behemoths across the horizon, it's gone. The loneliness, the emptiness, the absence of people, the absence of intrusions of people—all that would be lost."

Jacque Sundgren, a rancher along with her husband, Steve, once went door to door petitioning neighboring ranchers to say no to a proposed wind farm in their area. The project was never built. "This right here is my life," she said. "You put industry on it and it's gone." Steve explained the choice their family has made. "We could make hundreds of thousands of dollars off of wind, but there are other things to pass on to your kids."

It's no small wonder that Pete Ferrell has incurred the wrath of many Flint Hills ranchers and conservationists. Their passions and his pragmatism are a poor match. Pete's great-grandfather began assembling property for the family's ranch in 1888, with profits he earned from a dry goods store in Wichita. The family holdings grew to 7,000 acres, but Pete inherited only a third of the land. To keep the ranch intact, he bought back some of the acreage from his aunts and uncles. This left him carrying a crushing debt. "If you inherit land and it's debt-free, you can probably make a decent living, but paying interest and principal on land?"[14] The last time farmers and ranchers could break even on land that they had to purchase was in the 1920s, he tells me.

As he struggled to amass enough capital to purchase the remainder of the ranch, Pete eventually realized the toll this was taking on his health and well-being. "I was looking at a situation where I was going to join the Old Dead Ranchers Club, which is a group of great people who work 24/7, 365 days a year, until they fall over dead. 'Congratulations, you're a member,'" he says with a sardonic grin. We are sitting in his adobe-style ranch office, just across a rutted driveway from the hollow where he lives in a simple, unpainted wood cabin with his two dogs. The divorce from his wife some years ago, I can guess, only added to his financial stresses.

It was in the late 1980s that he became interested in the holistic management principles that economist Stan Parsons and ecologist Allan Savory had begun advancing through their teachings. Attending one of their training sessions, Pete was asked to catalog the full array of values on his property. He started underground with the oil resources that had yielded a handsome income for his mother over many years. (Even today, a half-dozen of those wells remain visible on his land, although Pete says they're drying up.) Then, rising to the surface, he reflected on the soils, the grass, the sun, and the water that make it possible to graze his cattle. Finally he was asked to delete those activities that are unsustainable or are not ecologically sound. Wind was on the short list of what was left.

Half-obscuring the windows in Pete's office today are two NREL maps showing the distribution of wind resources in Kansas. Even without those maps, he knew that the winds in his section of the Flint Hills were exceptional. This was confirmed when, in 1994, a representative of Oxbow Power Corporation knocked on his door, asking if he'd be interested in leasing out his land for a wind farm.

Pete found it strange that Oxbow would be interested in wind. The Palm Beach–based company was owned by Bill Koch, part-owner of Wichita-based Koch Industries until he split acrimoniously from the family-run conglomerate in 1983 and formed his own array of businesses with a heavy focus on traditional energy sources: coal, natural gas, and coke, an oil-refining byproduct used to power cement kilns and other industrial facilities. Pete also wasn't convinced that wind would be right for his ranch. In an effort to persuade him, Oxbow flew him out to California to visit some wind farms. There and elsewhere, Pete spoke with a number of wind farm hosts to see how the turbines affected their agricultural operations. They told him: "What wind turbines? We don't even think about them anymore. We still farm or ranch just like we always did."

Reassured, Pete signed a thirty-five-year lease with Oxbow, giving it the option to build a wind farm on his property. Meteorological towers then went up, but Oxbow came back to him in 1998 with discouraging news. "You have one of the finest wind resources we've ever seen," he was told, "but we're pulling out of here." Oxbow's agent referred to the project's unfavorable economics, but he also mentioned that a hostile political environment in Kansas contributed to the decision.

It wasn't clear to Pete whether these hostilities were at the state level, in the Flint Hills, or among the notoriously feuding Koch brothers.[15] Whatever the immediate reasons for Oxbow's abandonment of Flint Hills wind development, it is a sad irony that Bill Koch went from wind energy prospecting in Kansas to bankrolling anti-wind advocacy in Massachusetts. An avid sailor with a vacation home on Cape Cod, Koch—a conservative Republican—found common cause with liberal icon Ted Kennedy in fighting Cape Wind. *Forbes* magazine reported that, by the fall of 2006, Koch had donated $1.5 million to the nonprofit Alliance to Protect Nantucket Sound, orchestrator of the multiyear campaign against the offshore wind project.[16] He was the organization's co-chair at the time. Through Oxbow, he contributed another $620,000 to lobbyists who worked the halls of Congress in 2006 and 2007 on the anti–Cape Wind amendments to the Coast Guard authorization bill and other measures aimed at stopping the project.[17] Bill Koch may have estranged himself from Koch Industries, but his railing against Cape Wind was right in step with the far-right political activism of his brothers Charles and David, who have contributed well over $50 million to studies disputing climate change science, to policy initiatives opposed to curbing U.S. greenhouse gas emissions, and to politicians willing to fight for oil and gas industry interests in Congress.[18]

While Koch was navigating a quick course away from Flint Hills wind, Pete Ferrell continued collecting wind data from the meteorological towers Oxbow left on his ranch, and he began shopping for another developer. He ended up with seven offers and chose a small company called Greenlight Energy Resources, out of Charlottesville, Virginia. He had met the company's CEO, Sandy Reisky, on a bus tour of the Gray County wind farm in southwest Kansas, and the two immediately hit it off, signing a deal in 2001.

Then came the real shock for Pete. He knew that some of his neighbors had their misgivings about wind turbines in the Flint Hills, but he assumed they would come to see its benefits once they learned more about the technology, just as he had. Looking back on the lawsuits, the fiery public hearings, and the angry words from people he once considered friends, he now realizes how naive he had been. "We had TV cameras in the courtrooms, we had people pounding podiums and shaking their fists," he recalls with obvious sadness. "You think abortion is a hot issue, you just try to build a wind farm in the Flint Hills."

The Butler County Commission approved Greenlight Energy's permit application in 2003, but by then the outcry against wind energy had reached such a feverish pitch that Governor Kathleen Sebelius intervened. Mindful that permitting authority resided with counties rather than the state, she convened a citizen task force and instructed its members to recommend voluntary guidelines that counties might follow in balancing wind energy development with prairie preservation. Rose Bacon was appointed to this panel. "I think I was chosen because they didn't know me," she told me. "As a ranch wife, what possible harm could I do?"

The governor had hoped that consensual guidelines would emerge, but that was not in the cards. Although the task force's scope was statewide, the real friction points were in the Flint Hills. One faction, which Rose considered pro-wind, called for a three-tier classification system that would divide native grasslands into no-development zones, areas with restricted development, and areas with few restrictions. These zones, advisory in nature, were to leave the ultimate decision making with municipal and county authorities. Rose Bacon's faction took a much harder line, calling for an outright ban to be imposed by the state on wind energy projects in all areas with largely intact prairie, including a seven-mile buffer zone around those areas. It also recommended that the state's property tax exemption on wind energy investments be abolished, and favored consideration of a new statewide wind-development impact fee.[19]

Faced with this stalemate and sensing the growing vehemence of anti-wind forces in the Flint Hills, Governor Sebelius came up with an artful compromise in November 2004. On one hand, she asked—but did not order—wind developers with project proposals in an area she called the "Heart of the Flint Hills" to freeze their projects so that counties could have the time to prepare their own guidelines for wind development. The designation covered about two thirds of the Flint Hills—far less than Rose Bacon and her faction had sought—but the governor's action validated the Flint Hills defenders' claim to a unique historic landscape.

At the same time, the governor—a supporter of wind energy for Kansas—encouraged wind developers to build in other parts of the state, just as the Zilkhas decided to do when opposition mounted against their proposed project in the Flint Hills.[20] As Greenlight Energy's wind farm site lay just a few miles outside the Heart of the Flint Hills' southern boundary, Sandy Reisky and Pete Ferrell moved ahead with their project. To raise sufficient capital for the wind farm, Reisky sold the project to PPM Energy of Portland, Oregon, in December

2004, which later became a subsidiary of the wind development giant Iberdrola Renewables. Construction began early in 2005, and the 150-megawatt Elk River Wind Power Project went into operation in December of that year.

Sebelius never converted her Heart of the Flint Hills freeze to an outright ban on wind development in the area, but Rose Bacon and other ranchers formed a group, Protect the Flint Hills, to lobby for a stronger state policy. At the same time, they worked on fellow ranchers to dissuade them from signing any new deals with wind developers. Rose's advice was unequivocal. "If you care about your land, you won't lease it to wind. It's that simple."

In May 2011, anti-wind activists in the Flint Hills achieved another victory when Governor Sam Brownback announced a Road Map for Wind Energy Policy that discouraged any new wind development in an area more than twice the size of Sebelius's Heart of the Flint Hills. Though still not a categorical prohibition, the policy made it clear that wind developers would have a hard time building new projects in the 11,000-square-mile area that he dubbed the "Tallgrass Heartland."[21] Rose was elated. "Ten years of work have gone into this, and we finally have a definitive decision on the future of the Flint Hills," she wrote me. "It's green and beautiful here, the cattle are out to grass, and today this is the BEST place in the world to be!"[22]

The Elk River wind farm, with half its turbines on Pete Ferrell's property, falls inside the newly expanded boundaries of the Tallgrass Heartland. While expanding that project is now out of the question, Brownback has offered assurances that existing wind energy production in the Flint Hills will not be shut down. The governor also made it clear that transmission lines will not be blocked from crossing the Flint Hills. This comes as relief to wind developers in central and western Kansas, as well as points farther west, who are concerned about getting their power to eastern markets.

Despite all the turmoil, Pete's enthusiasm for wind energy has by no means abated. Along with stabilizing his ranch's finances, he firmly believes he is helping wean America off energy resources that, in the greater scheme of history, are moments away from running out.

To make this point, he unfurls a few dozen feet of computer paper, starting on the floor of his office, continuing out into the foyer, and then reaching across the full length of a second room where his ranch manager keeps track of cattle shipments. This long paper trail, ragged from repeated use, likens the past 750 million years of history to a single calendar year. Pete takes me to the final few perforated sheets and points to December 11, the date when mammals show up. *Homo erectus* appears the day after Christmas; *Homo sapiens* on December 30. At three seconds to midnight on December 31, we begin to burn petroleum, and by three seconds into the New Year, Pete tells me it will all be gone. "We have a whole civilization built on something that's gonna run out," he told a roomful of fellow Grinnell alumni when he lectured on this topic at a recent college reunion.[23]

Rose takes a cynical view of Pete's efforts. "People don't believe he's a wind conqueror; they believe it was for the money. . . . I feel bad that he had to make that choice because he's lost so much." I ask what he has lost, and she tells me: "Friendships and respect."

Pete, for his part, is now working on a project that may bring as much as 800 megawatts of wind power to cultivated farmland in southwestern Kansas—more than five times as much as Elk River Wind's installed capacity. Luckily for Pete and other Kansas wind developers, there are ample opportunities to tap the state's enormous wind potential outside the Flint Hills.

~

Grandpa's Knob, a few miles west of Rutland, Vermont, has a prominent place in the history of wind-generated electricity. There, on a 2,000-foot summit, an MIT-trained geologist-turned-energy innovator named Palmer Cosslett Putnam built America's first large-scale power-generating wind turbine in 1941. Equipped with two 70-foot-long aluminum alloy blades mounted on a 110-foot tower, the turbine was designed to produce up to 1.25 megawatts of power—much larger than the wind machines that were widely used in California forty years later and in the same league as today's commercial turbines. As protection against damage from rough weather, the blades could be pivoted so that their broad surfaces stood perpendicular to incoming winds. This safety measure, using a centrifugal or "flyball" governor, anticipated the more-sophisticated electronics of modern-day blade pitch controls.[24]

Putnam's giant wind machine had a short and sporadic set of runs. When the main bearing failed little more than a year after the turbine began operating, resource constraints imposed by the war effort kept it out of service for over two years. Then, only a month after it was finally repaired in March 1945, one of the giant blades broke off, shutting down the turbine once again.[25] In the immediate aftermath of World War II, one local company laid out tentative plans to rehabilitate the machine for power generation, proposing to make it a tourist magnet with a scenic toll road, summit restaurant, tower-top observation deck, and possibly a ski development.[26] None of that happened.

I climbed Grandpa's Knob in May 2010. To get there, I trudged up the same two-mile dirt road, steep and meandering, that had been used to truck turbine parts up to the site sixty years earlier. At the summit, all that remained of Putnam's wind machine were a few crumbling concrete footings and a small commemorative plaque. The site, however, did offer a panorama of three nearby ridgelines where a Vermont-based developer, financially backed by the Italian energy conglomerate Enel, had been negotiating to build a new 85-megawatt wind farm.

Leasing land for the wind farm turned out to be the easy part. The developer, Vermont Community Wind, had lined up more than 4,000 acres of woodlands managed by a large timber company, Wagner Forest Management. What proved much more difficult was getting residents of the affected communities to come to terms with a wind farm in their midst. The town of Ira was a case in point. With only 460 residents, the town stood to gain $10,000 to $11,000 per turbine per year in contributions from the wind developer. Given plans for multiple turbines within Ira's boundaries, total payments would have easily exceeded the town's $160,000 annual budget.[27] Yet this promised financial boon wasn't enough to overcome local residents' concerns about planting turbines on the forested ridges above their homes. Vermont Community Wind had originally identified sixty potential sites for turbines in five area towns. By June 2009, public opposition had reduced that number to forty-five, and by January 2010, the company had brought the total down to thirty-four.[28]

Even at this smaller scale, the wind farm had many local residents up in arms, not only about viewshed impacts, but also because of the noise that they feared the turbines would bring into their lives. Mary Pernal lives in a modest house in the town of Poultney, 500 feet below one of the ridges that Vermont Community Wind had targeted for turbines. An assistant professor of English at Green Mountain College, she described her apprehensions: "Most of us live in this neighborhood because it's a wild, serene, beautiful place, and the noises that we hear are pleasant noises: the sound of a creek gurgling by us, the sounds of birds, the sounds of various wildlife occasionally, the sounds of rain on the roof." She worried that the thirteen proposed turbines would create an amphitheater effect, filling her tranquil surroundings with unwelcome noise.[29]

Pernal and her neighbors were not just idly imagining the noise that wind turbines might bring into their lives. They had read or heard accounts of other New England residents living near wind farms—people like Phil Bloomstein, neighbor to the Beaver Ridge Wind Project in Freedom, Maine, who reported that noise produced by a GE 1.5-megawatt turbine 1,000 feet away from his home "can turn from an almost tolerable drone to a pulsating nightmare so oppressive that any outdoor activity is challenging."[30] Or Wendy Todd, who testified before the Maine Governor's Task Force on Wind Power that there's no escaping the noise and vibrations coming from the Mars Hill wind farm, 2,600 feet from her home in Aroostook County. Todd likened the noise from these turbines—the same GE model used at Beaver Ridge—to "a fleet of planes that are approaching but never arrive," and described particularly bad periods when "a repetitive, pulsating, thumping noise . . . can go on for hours or even days."[31]

Climbing Grandpa's Knob was an intriguing hike into wind energy history, but my primary purpose in visiting Rutland was to get a better sense of the mounting controversy about wind farm noise. In its Winter 2010 *Heart Health* newsletter, the Rutland Regional Medical Center featured one of its senior cardiologists linking wind turbine noise to sleep deprivation, which he said can cause an elevated risk of hypertension, heart attacks, atrial fibrillation, and stroke. The article angered a number of readers, who objected to the physician's speculative leaps and protested the medical center's apparent taking of sides in the stormy local debate about wind energy.[32] In a postcard sent to all its readers, the medical center apologized for the communication, declared itself to have no position on wind energy, and announced that it would be holding a public forum on the technology's health effects.

In late April 2010, just about a week before the forum was to be held, Vermont Community Wind announced that it was freezing its plans for the wind farm. Spokesman Jeff Wennberg was candid about the company's frustrations. "Vermont seems to say all the right things in terms of incentives and general state policy," he told a Vermont Public Radio reporter, "but when you actually get to proposing specific projects, it's very clear that the words and the reality of what is possible to do in Vermont [are] very, very different."[33]

Even with the wind farm on indefinite hold, close to a hundred people crowded into the basement lecture hall at Rutland Medical Center. The forum featured a debate between Dr. Robert McCunney, an internist specializing in occupational and environmental health at Massachusetts General Hospital, and Dr. Michael Nissenbaum, a radiologist at Northern Maine Medical Center who has become a self-taught expert on the noise impacts of wind energy. McCunney had recently co-authored an expert review of wind turbine sound and health effects for the American Wind Energy Association (AWEA) and its Canadian counterpart.[34] Nissenbaum had coauthored a searing critique of that study for an anti-wind group called the Society for Wind Vigilance.[35]

McCunney started out by describing the two major components of sound: frequency, measured in hertz, or cycles per second; and volume or loudness, measured in decibels. Wind turbines produce a variety of sounds: mid- to high-frequency aerodynamic sounds given off by rotating blades; lower-frequency sounds emitted by turbine machinery such as gears and generators; and ultra-low-frequency "infrasound" sometimes associated with turbine operations. High-frequency sounds, though audible at lower volumes than low-frequency sounds, attenuate more quickly over distance and are absorbed more readily by landscape features such as trees and leafy crops. Infrasound isn't even detectable to the human ear at low volumes, but it can be experienced as vibration and can create secondary vibrations or rattles in nearby structures.

Nissenbaum called the mélange of sounds produced by wind turbines "an acoustic pizza." Some mechanical sounds are constant; other sounds pulsate, like the swishing of blades as they cut through the air at tip speeds that can reach 200 miles per hour. Some are tonal; others are not. When multiple turbines are within range, as is often the case at commercial wind farms, they produce even more complex combinations of sound.

The two physicians agreed that, above certain thresholds, the sounds generated by wind farms become unwanted noise. Reactions to this noise vary, said McCunney. "Some people may be annoyed being stuck in a traffic jam, or being stuck too long in a line at the post office, or waiting too long for [their] son to come home at two a.m. on a Sunday night," he observed. Likewise, McCunney said, some people are more bothered by wind turbine noise than others. He referred to a Dutch study's finding that a small percentage of people—about 5 percent of a sample of 725 people living within 2.5 kilometers (4 miles) of a wind turbine—reported being annoyed by wind turbine sounds in the 35- to 40-decibel range. Two separate Swedish studies found that, when turbine sounds reached 40 to 45 decibels, 18 percent of respondents registered annoyance.[36]

Nissenbaum shared McCunney's view that people have very different thresholds for perceiving and tolerating noise. "One person in seven in this room will sense noise at six decibels lower than the average person," he said. "We're all built differently." Where he parted ways with McCunney was over the health impacts of wind farm–generated noise. McCunney insisted that there have been no rigorous studies showing a risk of adverse health effects from wind turbine noise. Nissenbaum countered with findings of his own limited investigation at the Mars Hill wind farm. In his evaluation of 22 people living within 3,500 feet of the turbines, 18 reported new or worsened sleep deprivation, 17 said they experienced persistent anger, 9 said they newly suffered from chronic headaches, and 8 had new or worsened depression. People living three or more miles away reported dramatically lower incidences of all these symptoms.[37]

Systematic studies have yet to explore the health effects of wind turbine noise on broad populations, but anecdotal accounts link a range of ailments to wind turbine noise. The U.S.-based Industrial Wind Action Group, whose declared mission is to "counteract the misleading information promulgated by the wind energy industry and various environmental groups," has published numerous testimonials by aggrieved wind farm neighbors on its website.[38] In Australia, health complaints by citizens led the government's National Health and Medical Research Council to evaluate their grievances. Based on this review, it stated: "While a range of effects such as annoyance, anxiety, hearing loss, and interference with sleep, speech and learning have been reported anecdotally, there is not published scientific evidence to support these adverse effects of wind turbines on health."[39]

Physician Nina Pierpont has coined the term "Wind Turbine Syndrome" as a catchall category for the health effects that her own anecdotal research has associated with wind turbine noise. In her interviews with ten families living near commercial-scale wind turbines, she heard about symptoms including sleep disturbance, headache, tinnitus (ringing of the ears), dizziness, nausea, visual blurring, tachycardia, irritability, concentration and memory problems, and panic episodes. Many of these conditions, Pierpont hypothesizes, occur when low-frequency sounds upset the normal balancing or "vestibular" function of the inner ear, as well as the brain's ability to process balance-related neural signals. She further hypothesizes that turbine-generated vibrations interfere with the gravity receptors in our visceral organs. These disruptions, taken together, create what she calls "Visceral Vibratory Vestibular Disturbance."[40]

Pierpont's study has been roundly attacked for its methodological flaws, including the lack of a control group living away from wind turbines that could provide a baseline against which Pierpont's reported health effects could be compared. Geoff Leventhall, a British acoustical expert, has taken Pierpont to task for her tenuous assumptions linking turbine-generated infrasound to ailments previously associated with much higher noise exposures by workers in aerospace and other heavy industries.[41]

Pierpont, in turn, has scolded critics like Leventhall for overstepping their professional bounds. "Deciding whether people have significant symptoms is not within the expertise of engineers or specialists in acoustics," she caustically noted in one of her talks. "[T]he hallmark of a good doctor is one who takes symptoms seriously and pursues them until they are understood (and ameliorated)."[42]

While Pierpont's findings may be sketchy, enough people are upset by wind farm noise to create real problems for an industry that seeks to cover a lot of new territory, quite literally, in the coming years. Even if we set aside speculation about various diseases arising from turbine noise, an increasingly vocal cohort of angry, sleep-deprived people will not be good ambassadors for an emerging industry that needs to win the public's support. Along with better research into health effects, what's needed are measurable and enforceable limits on noise. Also needed are siting guidelines that ensure neighbors a sufficiently generous setback from turbines to protect them from unwanted noise and other annoyances such as the pulsating light, called "shadow flicker," that rotating blades can cast across the landscape when the sun is at a low angle on the far side of a turbine.

To set noise limits for wind farms, a decision has to be made about what level of turbine-generated noise is acceptable. Acoustic experts often approach this challenge by comparing wind turbine noise to other, more familiar sounds affecting our daily lives. A dishwasher in the next room produces about 50 decibels of noise. A library interior or a suburban area outdoors at nighttime might register 40 decibels—

which is perceived to be half as loud as 50 decibels given the logarithmic scale of sound measurement. Sound at 30 decibels—half as loud again—would approximate a quiet bedroom at night or a quiet rural area with no wind, insects, or traffic.[43]

Today there are no federal noise limits for wind turbines, but the U.S. Environmental Protection Agency (EPA) generally recommends that outdoor noise levels should be no higher than 55 decibels during the day and 45 decibels at night.[44] The World Health Organization (WHO)'s Guidelines for Community Noise, issued in 1999, also call for a 45-decibel nighttime limit, measured outside the home on the assumption that noises in a bedroom with the window slightly open would be 15 decibels below that level. In an updated report focusing on Europe, the WHO has recommended a more-protective 40-decibel outdoor nighttime limit. Preventing sleep disturbance is the primary goal of this guideline.[45]

In the absence of enforceable federal requirements, some states have adopted their own general noise standards. Maine, for example, has set general noise limits that mirror the EPA's daytime and nighttime guidelines.[46] Most states, though, have treated noise as a matter to be governed by counties and municipalities. In some local jurisdictions, new ordinances specifically regulate wind farm noise.[47] Others have deferred to individual wind developers, letting them negotiate ad hoc arrangements with neighbors at their project sites.

Setting numerical limits for wind farm noise is complicated by the difficulty in capturing the *quality* of the noise generated by turbines, not just the *quantity* of that noise in decibels averaged over a certain period. Some early-design turbines with rotors mounted downwind of towers were known to make a distinctive thumping noise when their blades passed by the tower. The switch to upwind rotors has reportedly reduced this problem, although even today wind farm neighbors like Phil Bloomstein and Wendy Todd are annoyed by the pulsating sound that sometimes comes from the upwind turbines operating near their Maine homes. This noise apparently results from sound pressure that builds up during the blade's downward sweep.[48] If a three-bladed turbine is rotating at fifteen revolutions per minute, aerodynamic noise can be expected to spike about once every 1.3 seconds. At twenty revolutions per minute, it peaks every second. Noise from multiple turbines can complicate the sound rhythm further, causing wind critics like Mike Nissenbaum to bridle when turbine-generated noise is likened to the steadier sound of a dishwasher or a refrigerator humming in an adjacent room.

Nina Pierpont challenges current turbine noise measurement techniques on another ground, taking issue with the commonly used A-weighted decibel scale, which reflects the human ear's sensitivity to different sound frequencies by underweighting inaudible sounds. Her contention is that this scale fails to capture the low-frequency and ultra-low-frequency sounds that may be having the greatest adverse health impact on wind turbine neighbors.[49]

Requiring turbines to be set back a minimum distance from nearby homes is another way to protect wind farm neighbors from noise. The advantage to relying on setbacks is their simplicity: compliance is easily measured, and ongoing enforcement isn't needed once a wind farm has been built. A very real downside, though, is their inaccuracy: setbacks are at best a crude proxy for numerical noise limits. Several environmental factors influence how far and in what direction turbine noise travels. People living downwind of turbines are more exposed to noise than those living on the upwind side. During temperature inversions, sounds stay closer to the ground and are audible at greater distances. Hard surfaces such as frozen ground reflect rather than absorb sound, causing it to carry farther. And the presence or absence of foliage can play a big role in how quickly sounds attenuate as they pass across cultivated fields, pastures, and woodlands.

Certain wind conditions can conspire with topography to increase wind noise. When the wind is blowing at blade height but ground-level air is still, turbine noise can be particularly noticeable. This is often the case at Mars Hill, Wendy Todd told the governor's task force. "There are many times when winds are high on the ridgeline but are near calm at our homes." On these occasions, she said, "It doesn't matter which room you go to, there is no escape from the noise."[50] Pete Ferrell told me he isn't bothered by turbine noise on his Flint Hills ranch, but he does hear the turbines when the winds don't reach down into the wooded hollow where his rustic cabin nestles.

The National Research Council reports that a commercial-scale turbine produces 90 to 105 decibels, with the sound attenuating to 50 to 60 decibels at a distance of 40 meters (131 feet) and 35 to 45 decibels at 300 meters (984 feet).[51] If these fairly crude estimates were right, a setback of 1,000 feet would provide reasonable noise protection to wind farm neighbors. However, a greater buffer is very likely needed to accommodate acoustical differences between turbines, daily and seasonal changes in weather, varying topography, and the heightened sensitivity of certain people to noise. Maine-based developer Rob Gardiner of Independence Wind believes that setbacks in the 2,500- to 3,000-foot range may be needed, especially in areas where turbines are sited on ridgelines. As for Maine's 45-decibel nighttime limit for turbines, he thinks it's sufficiently stringent. "All the problem sites in Maine have reported actual sound levels above that limit," he observes.[52]

Keeping turbines a half-mile or more away from homes may be what's needed, but such a broad margin of protection will not be an easy sell to many wind developers. Even before newly elected Wisconsin governor Scott Walker created a public uproar with his assault on the collective bargaining rights of labor unions, he angered wind developers when, in January 2011, he called for increasing his state's mandatory turbine setback to 1,800 feet from an adjacent landowner's property line.[53] Not only did wind proponents object to the length of the setback, but they rightly felt it was arbitrary to use the neighbor's property line, rather than distance from the nearest residence, as the delineation point if the primary goal was to protect people in their homes from noise. Michael Vickerman, executive director of RENEW Wisconsin, protested "the folly of Governor Walker's job-killing proposal" and warned that its adoption by the legislature would drive $1.8 billion in new wind power development out of state.[54] Vickerman and others favored the less-forbidding standard already adopted by the Public Service Commission, requiring the homes of wind farm neighbors to be set back by the lesser of 1,250 feet or 3.1

times the turbine's maximum blade-tip height.[55] In March 2010, the same legislature that approved the governor's union-busting agenda suspended the commission's wind siting rules, leaving wind developers in limbo and leading AWEA to describe Wisconsin as "closed for business."[56]

Drs. Nissenbaum and Pierpont's recommendations are far more restrictive than Governor Walker's controversial proposal. They have called for 2-kilometer (1.24-mile) setbacks from turbines in ordinary terrain, going up to 2 miles in mountainous areas.[57] While these two doctors' positions may lie beyond the outer fringes of what most regulators are demanding, their outreach to state legislatures and local planning bodies gives their message a prominence that the wind industry can ill afford to ignore.[58]

Commercial power production, by virtue of its scale, is an intrusive presence in our lives. Wind farms are hardly exempt from this liability, as I have learned in my travels over the past two years and from my earlier advocacy work in New England. While the industry will inevitably have its detractors, there certainly are ways to help wind farm developers reach a more harmonious coexistence with their neighbors. Decisions about siting wind farms in areas where feelings about natural, scenic, and historic values run high may be aided by more coherent siting guidelines that balance the protection of those values with the need to take much fuller advantage of our renewable energy resources. The Biodiversity Conservation Alliance, already discussed in chapter 7 as an outspoken player in Wyoming's sage grouse debate, has done a particularly impressive job weighing factors ranging from visual concerns to wildlife protection in its delineation of go and no-go areas for wind energy development.

Regarding turbine-generated noise, a more proactive governmental role is needed to reduce an obvious source of annoyance and discomfort. Current levels of exposure to wind farm noise may upset only a minority of wind farm neighbors today, but those numbers will grow as wind farms proliferate, particularly as we approach the Department of Energy's goal of having 20 percent of our power come from wind by 2030. The federal government could lead the way by adopting uniform noise standards and delegating enforcement authority to the states—just as it has done with many of our federal environmental, public health, and workplace safety laws. Short of setting national standards, the government could surely offer more coherent guidance to state and local agencies as they develop their own. The current regulatory vacuum is substantially responsible for the discontentment that is emerging among at least some people living near wind turbines. Whether disgruntled neighbors turn to the media or take their grievances to court, the wind industry is not being well served by the absence of an adequate, enforceable framework for developing and operating wind farms.

As with any new industry, knowledge about wind power's impacts is growing as the technology gains traction in the field. Wind energy's enormous promise as a cleaner energy resource will be realized more fully if new information about noise and other concerns is openly and rigorously evaluated and appropriate measures are taken to protect public health and well-being.

CHAPTER NINE

Greening the Grid

WHEN MERIDIAN WAY'S ROTORS are turning, electrons race down insulated cables to the base of each tower. From there, a network of collector lines gathers all the power produced by the wind farm's turbines and carries it via miles of underground collector lines to one of two transformer stations at the wind farm site. These transformers then boost the power to 230 kilovolts, readying it for dispatch to the grid. All of this happens almost instantaneously.

Horizon Wind is lucky to have two high-voltage transmission lines nearby. Running through the countryside on tall steel stanchions, these lines carry power to Empire District Electric and Westar, the two utilities that have bought nearly equal shares of Meridian Way's output. Brad Beecher, chief operating officer of Empire District Electric, leads me through what happens next. Knowing roughly how much power his company's 160,000 customers use, he has to be sure to feed an equivalent amount of electricity to the grid. "We know how much water our customers are taking from the bathtub and we have to put that much water in the bathtub," he explains metaphorically. Empire District's electricity comes from a variety of sources—mostly from coal and gas, although wind now supplies about 15 percent of its customers' needs. In addition to owning half of Meridian Way's output, the company has contracted for all the power coming from Elk River, the Flint Hills wind farm that includes Pete Ferrell's land.

Some of the electrons generated at Meridian Way may actually reach the homes, businesses, and factories of Empire District's customers, but electrons are notoriously promiscuous. "You can't tell which electrons flow to our customers' meters," Beecher tells me. Once on the grid, electrons intermingle, not just with other electrons generated by a single company but with all the electrons produced by all the power suppliers that are part of an area's wholesale power pool. These markets typically cover multistate areas and, in much of the country, are governed by entities known as regional transmission organizations or independent system operators. Empire District is part of the Southwest Power Pool, a consortium of power producers, transmission providers, and electricity distributors stretching across all or part of eight states. The Southwest Power Pool's origins go back to 1941, when a group of utilities marshaled all available electricity for wartime aluminum production in Arkansas.

With transmission lines readily accessible, the developers of Meridian Way and Elk River had a relatively easy time getting their power onto the grid. This is far from the case in many other wind-blown parts of the country. On the National Renewable Energy Laboratory (NREL)'s color-coded maps, America's richest wind resources run through stretches of the eastern Rockies and Great Plains where people are few and power demand is pretty stable. Tapping the winds that sweep through these areas is relatively easy. Finding a reliable market for the wind-generated electricity is the hard part. Because existing conventional power plants meet most local needs, wind developers must seek out more distant buyers, often in population centers many hundreds of miles away.

To get a firsthand look at some of the hurdles involved in moving wind-generated power from where it's produced to where it's most needed, I visited Wyoming, home to some of the nation's strongest and steadiest winds. Nowhere are those winds more robust than along a stretch of Interstate 25 that runs from Casper down to Cheyenne, in the southeastern quadrant of the state.

Bob Whitton stood with me on a wooden porch just off the kitchen of his modest wood-frame ranch house, about sixty miles north of Cheyenne and just a mile or so east of I-25. He pointed to the rolled bales of hay that block the lower half of the windows on the west side of his home. These makeshift barriers help deflect the winds that blow hard off the foothills of the Rockies before crossing the gently rolling grasslands where he raises a small herd of Black Angus cattle.

He bought his ranch when he retired from the Air Force in 1994, after thirty years of piloting F4s, F5s, and F15s.

Early one December morning not long ago, Bob gazed out his kitchen window off to the west. There, strewn along I-25 near the Bordeaux junction, he could make out the long rectangular shapes of three overturned semi-trailers, flipped during the night by the howling winds. Later he drove out to the highway and found three more semis lying on their sides.[1]

Bob's ranch is in an area that has earned NREL's top ranking, with average winds rated at over 10 meters per second, or 22.4 miles per hour.[2] In the cold winter months, they often reach 30 to 40 miles per hour, posing a particular danger to newborn calves. "The wind will kill a calf pretty fast; the cold, not quite so fast," he says. Shielding calves from the elements is one wintertime ordeal; keeping his herd fed is another, as hay rolled out for feed often gets carried away by the wind. "Every time the wind blows, it costs me a lot of money."

These winds can be a formidable burden, but Bob knows better than most how to translate them into an economic opportunity. Along with running his ranch, he chairs a coalition of pro-wind landowner associations called the Renewable Energy Alliance of Landowners (REAL). This innovative group has jettisoned the traditional model, whereby a wind developer scouts around for promising sites and then approaches landowners one by one, retaining the upper hand in lining up agreements for the lease of their property.

REAL's 300 members recognize that, along with owning lands that are good for cattle, they hold a valuable resource in the winds that rip through their section of the state. Instead of waiting for a company like Horizon or Invenergy to come to them, they have formed associations that actively seek out takers for the wind on their ranchlands. Some neighbors have created limited liability corporations; others work together informally. Some associations include as many as forty landowners; others involve just a few like-minded property owners. Together, REAL's members have 800,000 acres of southeastern Wyoming lands that they are ready to market for wind.

Back in 2008, Bob and some neighbors formed their own local

group, the Bordeaux Landowners Association, now one of REAL's member associations. Putting 15,000 acres on the table, they began reaching out to wind companies with a neat package that included wind data, topographical maps, photos, and other documentation. Their prospectus went out to about fifty developers, and several responded. Negotiations ensued, and in February 2010, they signed on with Pathfinder Renewable Energy LLC, a Wyoming company backed by Dallas-based Sammons Enterprises, a private holding company with close to $45 billion in assets.

I met Pathfinder's land agent Vic Garber at a public board meeting of the Wyoming Infrastructure Authority, a quasi-governmental agency that helps expand Wyoming's electric transmission infrastructure through planning assistance, financing, and even co-ownership of new lines and related facilities.[3] Vic described Pathfinder's total ambition for its wind farm: 2.1 gigawatts of installed capacity sited across roughly 100,000 acres of land on both sides of Interstate 25.[4] Built at that scale, it will dwarf all other wind farms operating in America today.

Within Wyoming, Pathfinder may ultimately be surpassed by another mega-project sited about 100 miles to the west. There, on a sprawling ranch covering 500 square miles of rugged upland territory, a local subsidiary of the Denver-based Anschutz Corporation is moving forward with a project that may harness as much as three gigawatts of Wyoming wind by 2015.[5] Philip F. Anschutz, the privately held company's owner, built his multibillion-dollar fortune on oil and gas exploration ventures inherited from his father. In the late 1980s, he began acquiring railroad interests and later diversified into telecommunications, digital video data management, and ownership of a nationwide network of movie theaters. Today he is part-owner of the Los Angeles Lakers and other sports teams. Wind is no impulsive dream for this hard-headed entrepreneur.

Recognizing the market potential for Wyoming wind in faraway places, public utilities and independent "merchant" developers are racing to build long-range transmission lines. One of these projects, the Zephyr, has as its starting point the village of Chugwater, just a few miles south of Bob Whitton's ranch. From there, it will travel due west into Idaho and then turn south toward Las Vegas, in the power-hungry Sun Belt. If all goes according to plan, this 1,100-mile electron expressway will be fully operational by 2016. Pathfinder Renewable Energy has committed to take two-thirds of the Zephyr's capacity; two other wind developers—Horizon and BP Wind Energy—have contracted for the remainder. The total project cost is estimated to be $3 billion, or a little less than $3 million per mile.[6]

Nevada and California are also prime markets for the Chokecherry–Sierra Madre wind complex. Anschutz is planning on building its own transmission line, the TransWest Express, picking up power from the company's 1,000 or so planned wind turbines in south-central Wyoming and carrying it directly to southern Nevada.

Some of the power moved by these two lines might be consumed in Nevada, which has a renewable electricity standard that calls for 25 percent of its power to come from renewable sources by 2025. The prize destination for Wyoming wind energy, though, is California, trumping all of its neighbors in both the scale of its power market and the ambition of its renewable energy mandate. California will need to import substantial amounts of power from out of state if it is to meet its 33-percent-by-2020 renewable electricity standard. In-state wind and solar projects just aren't being developed quickly enough to meet this mandate.

The Zephyr and TransWest Express, relying on extra-high voltage current, will be much more energy-conserving than the lower-voltage lines that carry Meridian Way's power to market. If the Zephyr, a 500-kilovolt line, were to rely on AC current, it would lose about 1.3 percent of its power every 100 miles. If it increased its voltage to 765 kilovolts, it would shed as little as a half-percent of its power over the same distance.[7] However, because both projects plan to rely on more highly efficient DC current, they may see as much as a 20 percent further decrease in their line losses.[8]

~

Building a major new power line is not simply a matter of choosing the right technology. The transmission developer must comply with an array of federal laws, including those that govern the use of federal lands and those that protect threatened and endangered species.

In Wyoming, as throughout much of the West, the federal government is a major stakeholder in the planning and siting of new transmission lines. One look at the pattern of land ownership in Wyoming makes it clear why. Nearly half the land in the state—about 30 million acres—is owned and controlled by the federal government.[9] Some of this land is in sprawling, contiguous parcels, while the rest follows a quirky pattern known as the Wyoming Checkerboard—a holdover from the 1860s, when Congress passed two successive acts aimed at opening up the West.[10] All along a rail right-of-way running across the state, alternating square-mile sections of land were handed over to the Union Pacific Railroad, extending out twenty miles on each side of the tracks.

Given the role of transmission in creating new frontiers for economic development, it's no coincidence that the labels given to several proposed interstate power lines resonate with the region's rail lore. TransCanada's Zephyr line shares its name with one of Amtrak's most dramatically scenic routes, rolling from Chicago across the Great Plains and the Rockies to San Francisco. The monikers of other planned lines similarly evoke the spirit of America's railroad heritage: the TransWest Express and High Plains Express, both running out of Wyoming, and the Green Power Express, rooted in the Dakotas.

The U.S. Bureau of Land Management (BLM), within the Department of the Interior, oversees a rigorous, multiyear process of evaluating transmission projects that cross federal lands. In doing so, it has to reckon with a range of factors, reflecting what one BLM official calls its "schizophrenic" mandate. On one hand, the agency is charged by various federal laws with promoting the exploitation of commodity resources on federal lands—underground minerals as well as surface assets such as timber and grasslands. On the other hand, another whole set of federal laws calls upon the BLM to conserve our natural resources for future use while protecting the wildlife and habitats on those lands.[11]

Amidst these mixed signals, the Energy Policy Act of 2005 pointed the BLM toward wind development with its call for at least 10 gigawatts of non-hydropower renewable energy projects to be built on public lands by 2015.[12] Tom Lahti, who is the BLM's Renewable Energy Chief in its Wyoming office, decided that the BLM should not simply wait for good projects to present themselves. Instead, he initiated a study to identify the most promising areas for wind development on Wyoming's federal lands. Access to present or proposed transmission lines is one priority. Another concern is preserving wildlife—making sure that key habitats for vulnerable species like sage grouse, eagles, and migrating bats are taken into account. Viewshed protection is a third consideration, giving special attention to the state's extensive network of National Historic Trails.[13]

Once a number of wind development areas have been identified, Tom expects that the BLM will lease them out on a competitive basis, just as it does today when offering large parcels of federal land for oil and gas drilling.[14] This would be fairer, he feels, than the BLM's current ad hoc approach, whereby parcels are leased out in response to wind developers' individual requests. Under the present system, Tom says, "The only competition is . . . who gets there first."[15]

The BLM and other federal agencies are also taking a proactive approach to charting out corridors that would allow a range of energy resources—oil, gas, and perhaps someday hydrogen, as well as electricity—to move more easily across public lands.[16] In November 2008, the government released its proposed West-Wide Energy Corridor, making it clear to transmission developers that, while they are not strictly confined to this corridor, it will be easier to gain federal approval for lines that stay within its alignment.[17]

Obtaining federal rights-of-way is only one piece of a complex set of negotiations that the developers of projects like the Zephyr and the TransWest Express must undertake. States vary widely in the degree of deference given to local zoning boards and county commissions. In Nevada, for example, counties and municipalities largely govern the siting of transmission lines,[18] whereas in Wyoming, primary control over transmission line authorization rests with two state agencies, the Industrial Siting Council and the Public Service Commission.

As it can take years to arrange rights-of-way over hundreds of miles of public and private lands, wind companies have to time their planning and development efforts carefully. On one hand, they don't want to have wind turbines sitting idle for months—or even years—while a transmission line is moving slowly toward completion. (This has been the situation in China, where wind developers have installed massive new generating capacity in Inner Mongolia and other rural provinces years before the State Grid has built the necessary transmission infrastructure.[19]) On the other hand, wind developers don't want to find themselves carrying the cost of reserved space on a transmission line that is ready for business long before their turbines are in the ground and the blades have started spinning.

Beyond federal and state agency review, proponents of new transmission lines must deal with the concerns of citizen groups, including those who suspect that certain of these lines may be stalking horses for coal power interests. Howard Learner heads up the Environmental Law and Policy Center, a multistate advocacy group headquartered in Chicago. His suspicions center on the Green Power Express, a 3,000-mile network of 765-kilovolt AC power lines, heralded by its proponent, ITC Transmission, as a $10- to $12-billion project "designed to efficiently move up to 12,000 megawatts of renewable energy in wind-rich areas to major Midwest load centers."[20] Howard directed me to a map on the ITC website, which shows the project's starting point in North Dakota's Antelope Valley. "The wind belt in North Dakota is not principally around Antelope Valley," he said flatly. "You've got to scratch the surface a little bit to figure out who the transmission is going to serve."[21]

I took Howard's advice and started scratching. Via the online watchdog group SourceWatch, I had no difficulty finding the Antelope Valley Station, an 870-megawatt coal-burning giant right where Howard said it would be. I also found another five coal-fired power plants within a forty-mile radius of Antelope Valley, adding up to more than 4,000 megawatts of capacity.[22] Might the Green Power Express end up carrying coal-generated electrons out of North Dakota, along with or instead of power from renewables? Nothing under federal law or policy would seem to bar this outcome. To the contrary, the Federal Energy Regulatory Commission (FERC)'s "open access" transmission policy, adopted in 1996, is designed explicitly to ensure open, non-discriminatory access to the nation's transmission system.[23]

Howard and his colleagues at the Environmental Law and Policy Center worry that lines like the Green Power Express could become part of a larger network of new transmission lines that deliver coal power, produced and sold relatively cheaply in the Midwest, to higher-priced electricity markets in the Northeast. The Union of Concerned Scientists, based in Cambridge, Massachusetts, shares this concern. In a 2008 study, the group looked at the market dynamics that could emerge as a result of the Regional Greenhouse Gas Initiative (RGGI), the Northeast's cap-and-trade program for carbon emissions. The study pointed to a critical flaw in the RGGI regime's focus on power plants in a single region. Conceptually, the auction-based trading of carbon allowances among northeastern utilities is intended to stimulate a switch to cleaner fuels and energy efficiency measures. In practice, the upward pressure that RGGI places on the price of coal-generated power could entice enterprising Midwest utilities to export their dirtier, cheaper power to the Northeast. Building a new cohort of power lines connecting the two regions would only make it easier for this to happen.[24]

Some East Coast politicians and energy planners express a very different aversion to new power lines coming out of the Midwest. They fear that stepped-up transmission would give wind power generated in the Midwest too ready a market in the East, where large-scale wind farms have been slow to emerge. Representative Ed Markey of Massachusetts, for decades a champion of renewable energy development, expressed this view when I visited his Capitol Hill office in April 2010. "The consensus in New England is to capture this wind revolution and make it our own," he told me, earning nods of approval from the three members of his staff who huddled with us around a coffee table strewn with books about U.S. energy policy. At the time, Markey held leadership positions on two key congressional energy committees—positions he lost when the Republicans gained a majority in the House in November 2010.

Markey's jealous guarding of East Coast wind was based on data he had seen that pointed to the region's surplus of untapped, primarily offshore wind. He said that 30 to 40 gigawatts of offshore wind was already economically developable. (That's about three to four times the total installed power-generating capacity in Massachusetts today.) Hundreds of additional gigawatts would be within reach once deepwater wind technology matures, he added.[25] This bullish embrace of offshore wind struck me as slightly odd, given the equivocal position Markey had long taken on Cape Wind during most of its tormented journey through state and federal permitting. It was not until November 2009, three months after Ted Kennedy's death, that Markey called explicitly for federal approval of the project. Presumably he had refrained from endorsing Cape Wind out of deference to Kennedy's outright hostility to wind turbines in Nantucket Sound. Few members of the Massachusetts congressional delegation had been willing to part ways with the state's senior senator on this controversial issue.[26]

FERC chairman Jon Wellinghoff takes a less-parochial stance on the nexus between wind and wires. He believes that wind could provide upwards of 50 percent of our power needs nationwide, and he is convinced that our most promising wind resources can only be accessed if we make an all-out commitment to building new transmission lines. He punctuates his public presentations with slides showing the advanced visions for grid development that are fast emerging overseas. On one slide, 25,000 miles of high-voltage conduits interconnect a future Europe with renewable resources ranging from Norwegian hydro to vast solar energy farms stretching across North Africa. On another slide, he shows China's fast-emerging supergrid, with row upon row of 800-kilovolt lines bringing power from wind-rich areas in the country's western reaches to major cities in the East.[27]

I asked the FERC chairman how he regarded environmentalists' concerns about coal piggybacking on new American power lines ostensibly built for wind. His response was blunt, even irritated: "I think it's an urban myth." Coal plants like those near Antelope Valley have all the transmission they need, he said, adding that new coal-fired facilities are unlikely to be built in the coming years, given uncertainties about the future price of carbon-based power production. As evidence of the slowdown in coal plant development, he pointed to the tilt toward wind in the commissioning of new generating capacity in 2009: "We had almost 10,000 megawatts of wind put on the system last year. That was almost ten times the amount of coal."[28]

A stepped-up investment in coal-fired power generation may not be as remote a possibility today as it was when I met with Wellinghoff in the spring of 2010. Carbon taxation and market-based trading of carbon allowances were prime targets of the anti-incumbent fervor that brought a new wave of Tea Party–inspired Republicans to Congress just a few months later. The American Clean Energy and Security Act of 2009, cosponsored by Markey and California congressman Henry Waxman, was unfortunately a lightning rod for this hostility. The nationwide cap-and-trade program that it proposed for carbon emissions from major industries was roundly attacked as a punitive drag on an already flagging economy. When the 112th Congress convened, cap-and-trade was dead. Also abandoned was Waxman and Markey's proposal for a nationwide electricity standard that would set explicit targets for power generation from renewable sources.[29] In its stead, discussions shifted to a much more amorphous "clean energy standard" that would include nuclear power, as well as advanced methods of burning coal. If this reformulation finds its way into law, there may be added cause for concern about the encroachment of nonrenewable electrons onto new "green" transmission lines.

Questions may remain about competing uses of an expanded and modernized grid, yet it's clear that without a major investment in new transmission, much of our nation's wind energy potential will remain beyond reach. Recognizing this, the drafters of the 2009 stimulus package allocated $6 billion in loan guarantees for renewable energy generation and electric transmission projects and another $4.5 billion in matching grants to modernize the grid.[30] In his Earth Day 2009 speech before wind tower factory workers in Newton, Iowa, President Obama was passionate in endorsing this investment. "The nation that leads the world in creating new sources of clean energy will be the nation that leads the twenty-first-century global economy," he declared, adding that "we also need a smarter, stronger electricity grid to carry that energy from one end of this country to the other."[31]

In fact, federal efforts to create priority transmission corridors for new energy investments predated the Obama administration's recovery program by several years. As early as 2002, the Department of Energy began exploring ways for FERC to break transmission bottlenecks when state agencies and regional planning bodies fail to move priority grid expansion projects forward.[32] The Energy Policy Act of 2005 then stretched FERC's role in transmission line siting, giving it authority to issue permits for new transmission investments in designated national priority corridors if state siting bodies "withheld approval" for more than a year.[33]

A firestorm ensued when FERC interpreted this law as giving it the power to step in where a state has *denied* a siting permit, as opposed to simply not acting on it.[34] The Virginia-based Piedmont Environmental Council, together with other citizen groups and state public service commissions, brought suit and won a federal circuit court ruling that preserved the prerogative of states to turn down new transmission lines. When the Supreme Court declined to review this ruling, it became clear that without new federal legislation, states would remain in firm control of transmission line siting.[35]

Today the debate continues to rage over who should be involved in siting decisions about new interstate transmission lines. Two leading industry groups, AWEA and the Solar Energy Industries Association, insist that FERC should have ultimate control over the siting of multistate transmission lines, similar to the authority it currently exercises over interstate natural gas pipelines.[36] Don Furman, a veteran wind developer who recently chaired AWEA's board, candidly refers to state officials as a mismatch for decision making on transmission investments. Governors and public utility commissioners have told him, "We're not getting paid to go out there and build major interregional transmission. It needs to be done, but you ought not ask us to do it, to approve it, because that's not our job."

Furman likens the U.S. grid to a patchwork of country roads. What's needed, he says, is equivalent to our interstate highway system. Just as the interstate's planners and builders relied heavily on the federal government's strong guiding hand, he favors firm federal leadership in building a superhighway network for electric power.[37]

Howard Learner remains wary of giving FERC too free a hand in authorizing new transmission lines. Although his group is an active proponent of wind power, the Environmental Law and Policy Center has also worked to strengthen and enforce state laws protecting wetlands and other natural habitats. These laws, he fears, could be compromised if their protections are downplayed or preempted by a new federal siting regime.

Chris Miller of the Piedmont Environmental Council is even more adamant about keeping the federal government away from transmission-line siting. Having won the court battle that now limits FERC's power to override state siting decisions, he strongly rejects the view that the global climate crisis calls for decisions that could trump local planning priorities. "I don't see climate change and the international consensus of scientists as being any more politically valid than the desire of people here to have control over their viewshed," he tells me. At some level, he acknowledges that climate change is a problem. To address it, he argues half-heartedly for energy efficiency and greater use of passive solar building design. He concedes that these steps will fall far short of getting us the cutbacks in U.S. carbon emissions that so much of the scientific community is calling for, but he resents the "energy geeks" who are pressing for deeper cuts in pursuit of what he calls "an unratified policy of 80 percent greenhouse gas reduction."[38]

I met with Chris at the Piedmont Environmental Council's headquarters, in a beautifully restored clapboard house just off Main Street in historic Warrenton, Virginia. Driving into Warrenton just before our meeting, I grasped just how far area planners are willing to go to preserve at least the surface trappings of colonial authenticity. Tract houses spill across the rolling hills, their fake clapboard aluminum siding only slightly more convincing than their make-believe window mullions. Post-and-rail fences made of white molded plastic only add to the effect, bracketing the road, dipping down into ravines, and running across empty expanses of close-cropped grass. This is horse country minus the horses.

As I drove out of Warrenton past shopping malls and medical offices designed to look like eighteenth-century brick manor houses, I couldn't help thinking how tough a time transmission planners will have building a green-power superhighway worthy of the name. However the balance is struck between state and federal decision making on transmission-line siting, tensions between widely divergent values and worldviews will only grow as we move from the abstract notion of a twenty-first-century supergrid to the reality of laying new wires across the American landscape.

To supply 20 percent of our electricity from wind by 2030, the Department of Energy estimates that we will need to invest about $60 billion in expanding the American grid.[39] That would amount to about $30 per household per year, or a monthly cost of $2.50 per household, between 2012 and 2030. While this may sound like a very modest price for a monumental resource shift in our electricity sector, the dissension over who should pay for new transmission lines has been almost as explosive as the jurisdictional debate over siting them. FERC has refrained from imposing a one-size-fits-all allocation formula on these new transmission investments, instead inviting multistate transmission groups to develop schemes that fairly reflect the range of generation needs and ratepayer benefits in their service areas.

One regional transmission organization, the PJM Interconnection, coordinates the flow of power across all or part of thirteen Midwest and Mid-Atlantic states and the District of Columbia. More than 160 gigawatts of generating capacity and more than 50,000 miles of transmission lines are under PJM's control. When PJM proposed to spread the costs of major new transmission lines evenly among all utilities in its system, it raised the hackles of Midwest utilities that rely mainly on lower-voltage power lines to deliver electricity over relatively short distances.[40] The power companies claimed that PJM's cost-spreading formula would force them to underwrite transmission lines built primarily for the benefit of electricity consumers in the eastern PJM states.

The Southwest Power Pool has just adopted a more nuanced scheme for spreading the costs of new high-voltage transmission across its eight-state region. This "Highway-Byway" formula assigns the full costs of "Electricity Highways"—high-voltage, longer-distance lines—to all system users. "Electricity Byways"—mid-voltage lines generally covering less terrain—allocate a third of costs on a systemwide basis, leaving two-thirds to be paid within the zone where the project is located. And all the costs of lower-voltage lines are borne locally.[41]

Although it's an unlikely stretch given the prevailing antitax sentiments in Congress, some have proposed a new federal transmission tariff to supplement regional cost recovery mechanisms like the Highway/Byway formula. There is precedent for such a tax in the federal portion of the gasoline tax, which has been a financial mainstay of the interstate highway system since the 1950s and, to a lesser extent, a revenue source for federally supported mass transit projects. Susan Tierney, who headed up the Department of Energy's policy efforts during Bill Clinton's administration, symbolically invokes the federal highway system when she calls for this tariff to support a nationwide "interstate electric highway system." She asserts: "State and utility-service-territory boundaries have no more meaning for the transmission grid than they do for the transportation highway system, since electrons flow across boundaries according to the laws of physics rather than the laws of states." More practically, she justifies this investment because of the essential national purposes it would serve: the provision of a "reliable, economic, secure supply" of power and the shift toward a U.S. economy based on clean, renewable energy.[42]

However the costs of new transmission lines are shared, there is no escaping the fact that building new power conduits across large expanses of the United States will come at a considerable cost to the American public. Arriving at a fair way to socialize these costs will be crucial to gaining public acceptance of a new energy infrastructure that will give wind and other dispersed renewables their rightful place in our energy mix.

Building a wind-friendly transmission system is not just about stringing up wires and finding ways to pay for them. Accommodating the wind's variability demands constant vigilance. When winds drop off unexpectedly, grid operators must act immediately to make sure that electricity customers don't suddenly find themselves without electric power. When wind-generated output rises above predicted levels, system operators must take steps to absorb that power without destabilizing the grid.

Hydropower has long been used to help overstressed power systems cope with periods of peak demand, particularly on hot summer days when air conditioning needs outstrip the capacity of baseload-serving fossil fuel and nuclear power plants. As wind energy has become a more significant power resource in recent years, hydro dams and pumped-storage reservoirs have increasingly been called into play to keep power systems in balance. Within minutes, a hydro facility can step up its power output by channeling more water through its turbines. To soak up excess power, it can activate pumps that will transport water to an elevated storage reservoir. Coal and nuclear plants, which are slow to heat up and cool down, simply cannot make the swift shifts in power often needed to work in tandem with variable winds.

While hydropower and quick-starting, gas-fired turbines play an important role in keeping the grid balanced, today's transmission innovators are coming up with a variety of new technologies and management tools. Jon Wellinghoff brims with enthusiasm when he talks about innovations like smart-metered electric vehicles that allow a centralized, automated dispatcher to send incremental pulses of power to and from plugged-in car batteries to help maintain a balance between energy demand and available electric current. Most cars, he says, sit idle for twenty to twenty-two hours a day. Connected to the grid during those many hours, tens of millions of vehicles in an electric car fleet can serve as a vast balancing resource, easily accessed by grid operators. He and Knud Pedersen of Denmark's DONG Energy would have a lot to discuss as Denmark moves forward with its own electric vehicle network.[43]

Wellinghoff also gets excited about buildings whose heating and cooling can be controlled remotely to ease power use in low-wind periods and increase consumption when the winds are strong. In the same animated way, he speaks of "sentient" appliances like refrigerators with multiple functions—defrosting, ice making, moisture control, and ordinary cooling—programmed independently to modulate power use. "You don't care when your refrigerator compressor is on. You don't care when your defrost cycle is on," he explains. "You only care that you have a cold beer, nothing defrosts, and you have ice when you need it."[44]

Utility-scale flywheels are now used in a few areas to even out bumps in power flow from second to second and minute to minute. This service is called "frequency regulation." Wellinghoff, a longtime Nevadan, tells me that my home state of Massachusetts is at the forefront in applying this technology. In Tyngsboro, the Beacon Power Corporation has installed 3 megawatts of flywheel storage and is on its way to integrating multiple flywheel units into a 20-megawatt storage system. These devices store excess power by converting it into kinetic energy: a carbon-fiber cylinder, vacuum-sealed and levitated by magnetic bearings to minimize friction, rotates at speeds of up to 16,000 revolutions per minute. When power is needed, the flywheel's stored energy is tapped to drive a generator that feeds electrons back into the grid. Flywheels are sometimes called grid "shock absorbers" because of their nimbleness in taking on and discharging energy in rapid response to grid-balancing needs.[45]

Flywheels may be nimble, but Throop Wilder maintains that his company's batteries will do a better job matching the variability of wind power over multihour periods. Wilder is president of 24M, a Massachusetts-based startup located in a small suite of offices adjacent to the MIT campus in Cambridge. According to Wilder, the 24M battery, which uses a lithium ion–based semisolid suspension, will bring down the cost of batteries and reduce the space they require. "We will be able to deliver 1 megawatt of power over a four-hour period from a battery the size of a small walk-in closet," he claims. This compactness will allow battery storage to be sited in urban locations, right near consumers, rather than relying on remote multi-acre sites now used for pumped hydro storage reservoirs—a current mainstay in balancing power from multiple generation sources.[46]

Underground injection of compressed air is being explored as another means of energy storage that could help level out the fluctuations in power produced by wind farms. With this technology, surplus electricity is used to compress air, which is then pumped into a geological formation such as an abandoned mine, a salt dome, or a capped layer of subterranean sandstone. When power is needed, the air is brought back to the surface. As the air is reheated, its expansion creates sufficient pressure to turn a set of turbines. In Iowa, a consortium of municipalities spent several years studying a thousand-acre site where air would be pumped into a sandstone aquifer two thousand feet below ground. One of the state's top proponents of community-based wind power, appropriately enough named Tom Wind, was a lead consultant on the project. Though this site was ultimately determined a poor match for the planned 270-megawatt project, two other utility-scale compressed-air energy storage facilities are already operating—in Alabama and Germany.[47]

As we work to green the American grid, we need new planning tools, siting provisions, and financing strategies that will help us create a grid that measures up to the extraordinary wind resources across our continent and off our shores. Regional transmission organizations have a vital role to play in coordinating current grid operations, assessing future needs, and coming up with fair and equitable cost-sharing mechanisms for new projects. Ultimately, though, strategic investments in new grid infrastructure may require a higher degree of national direction to break through state and local logjams.

Building a twenty-first-century grid invites ingenuity as well as collaboration. From Beacon Power's flywheels to 24M's batteries, smart innovators are already rising to the challenge. The groundwork for a new power infrastructure that makes wind a major player is now being laid. The next steps will require clear vision and careful planning, as well as political leaders who are committed to moving America to a new era of energy self-reliance and energy security.

Epilogue

NOW MORE THAN EVER, the winds of change are blowing. In the Middle East and North Africa, the upending of long-entrenched authoritarian regimes has laid bare just how much the United States relies on relationships of convenience with faraway countries whose political and social fabrics we scarcely understand. The need to free ourselves from an overwhelming dependence on energy resources outside our borders and beyond our control has never been more apparent. The OPEC oil embargo of 1973 stirred a new awareness of the need to wean ourselves off foreign oil, but all too soon that clarion call was muffled by a more dominant message: petroleum was there for the taking, even if it threatened to draw us into one war after another in the Middle East.

On the home front, the opening of the western coal frontier and the exploitation of new natural gas reserves have given us the sense that we can fuel our electricity needs by burning through vast quantities of carbon, with little regard for the environmental consequences. Even though we are the world leader in per-capita greenhouse gas emissions, we have consistently balked at joining the Kyoto Protocol and other international efforts to avert global warming. The cost to our industries and consumers would be too high, our political and industrial leaders have insisted. Unfortunately, the same metric has not been applied to the fortune that we have spent over the past decade fighting ill-defined wars in the Middle East—more than a trillion dollars and counting.[1]

The Fukushima reactor disaster in March 2011 put us on renewed notice about the perils of nuclear energy just as the United States was considering a new wave of power plant construction. Instead of buffering Japan's bitter memories of Hiroshima and Nagasaki, the world's pursuit of civilian nuclear power has brought ever-greater numbers of nations to a roulette table where the stakes are impossibly high. We have begun to learn the hard way that rare events do happen—at Three Mile Island, Chernobyl, Fukushima, and who knows where next.

If wind energy looked promising in the first decade of this new millennium, it looks essential as we begin making our way through the second. Wind is no panacea, but the problems it poses are within our ability to address responsibly, and the scale of its potential contribution to our energy economy is staggering. By the end of 2012, we should have upwards of 50 gigawatts of installed capacity at wind farms scattered across most of our fifty states. That will get us a sixth of the way to the Department of Energy's goal of supplying a fifth of our power from wind by 2030, and at least start us down the path to FERC chairman Jon Wellinghoff's longer-term vision of having wind provide half of America's electricity. Happily, we have barely scratched the surface of the 11,000 gigawatts of wind power available to us on land. Equally wide open are the prospects for developing our wind energy resources at sea. Cape Wind, likely to become the nation's first offshore wind farm, has survived the obstacles thrown in its way and should be operational in a few years' time. That project will provide less than a half-gigawatt of installed wind power, leaving about 4,000 gigawatts of offshore wind resources for future development—far more electricity than America could conceivably need in the foreseeable future.

Even with our superabundance of available wind, it would be naive to believe that wind energy alone could free America from its reliance on fossil and nuclear fuels. Wind's variability requires balancing our wind-generated power with other technologies that are not constrained by intermittency. Realistically, fossil fuels like coal and gas, despite their drawbacks, will need to play a role in finding that balance during the next few decades. At the same time, we must make an all-out effort to develop a range of renewable technologies that can complement land-based wind power while doing less damage to the environment than fossil or nuclear fuels. Tidal and geothermal power, deepwater wind, carefully selected crop-derived fuels that have a net energy benefit and don't compete with our food production, and biogas drawn from sewage treatment, landfills, and livestock waste are among the options. Solar energy, intermittent like wind but with differently timed peaks and troughs, can further contribute to a balanced power supply, especially if its price becomes more competitive with other generation sources. The way forward will involve an integration of multiple technologies, with full recognition of the trade-offs involved with each.

Making wind a major American electricity provider will also require us to build a green-power superhighway worthy of the name. A new level of collaboration across state boundaries will be needed to plan these routes, and new mechanisms will have to be established to ensure that national energy priorities are given appropriate weight in decisions about new transmission lines. Beyond making sure that wind-generated power can be tapped where it is strongest and delivered to areas of greatest demand, sophisticated "smart grid" management tools and new power storage technologies will have to be developed to translate the wind's natural variability into an assured, continuous flow of power that perfectly matches consumer demand. The options abound, from flywheels and pumped hydro storage reservoirs to plug-in electric cars and industrial-scale batteries. All present timely opportunities to a U.S. technology sector that is looking for ways to make American industry more responsive to twenty-first-century needs.

To harness the wind's potential, we will need to sustain a level of federal support that places wind energy at least on a par with other new power plant investments. The U.S. government's stimulus program, with its flexible menu of federal production and investment tax credits convertible to treasury grants, has helped even out the playing field for wind energy in recent years. These programs were critical to averting a devastating setback to wind manufacturers and wind farm developers when the recession hit the industry hardest in 2010.[2] Continuing a program of wind energy tax credits will be a vital counterpoint to the subsidies, overt and hidden, that our government has long bestowed upon the fossil fuel and nuclear power industries.

Just as we dare not pin our hopes for wind on the elimination of government subsidies for fossil fuels and nuclear energy, we would be foolish to chart a course that depended on congressional passage of a federal tax on carbon emissions. More clearly within reach is a federal standard akin to the renewable electricity standards that have already been adopted by more than thirty states, requiring utilities to provide a minimum percentage of their power from renewable energy. Yet this approach, too, has proven vulnerable to lobbying attacks by the fossil fuel and nuclear industries. President Obama's commitment to wind, so persuasively articulated during his early days in office, yielded all too quickly to a more ambiguous embrace of "clean energy" that risks diverting us from renewable energy investments and sending us down a path toward "clean coal" and even "carbon-free nuclear energy," overlooking the horrendous hazards to health, safety, and the environment posed by each.

Even as our political leaders waffle on renewable energy's role in shaping our national energy portfolio, many of our states continue to press forward with policies that significantly advance wind and other renewable technology investments. With renewable electricity standards adopted by nearly two-thirds of our states, the momentum will continue to build. Texas, the nation's oil king, now generates nearly 10 percent of its electricity from wind. In Oregon, 15 percent of the power supply is already based on wind. California now draws nearly 20 percent of its electricity from renewable sources, and it is charting a course toward 33 percent reliance on wind and other renewables by 2020.

Leading proponents of renewable energy development in those and other states notably cut across party lines. Not only did Texas governor George W. Bush preside over the adoption of the state's first renewable electricity standard, but both Jerry Brown, California's current Democratic governor, and his Republican predecessor, Arnold Schwarzenegger, have carried the renewable energy banner with pride and determination. In November 2010, when the Koch brothers and other fossil fuel industry leaders bankrolled a referendum campaign to neutralize the renewable electricity standard and other elements of California's greenhouse gas agenda, one of the state's most vigorous defenders was George Shultz, Ronald Reagan's secretary of state and Richard Nixon's first budget director. Recalling the American car industry's shortsighted folly in opposing tighter air emission standards back in 1990, this lifelong conservative Republican observed: "There is a long history here of the pessimists underestimating what American ingenuity can do."[3]

The opportunities for ingenuity in developing America's wind energy potential abound. In my travels researching this book, I have witnessed this promise in the determined efforts of turbine manufacturer Clipper Wind. I have seen Rust Belt companies like Timken seize upon wind energy as its next big strategic bet. I have admired the pluck of family-run businesses like Cardinal Fastener as they carve out new niche markets. And I have been awed by the creativity of inventors and innovators who are racing to create a smarter grid.

But wind energy isn't just about innovation. It's about sound investment and careful planning by a growing corps of U.S. entrepreneurs. It's about creating a new generation of jobs in a renewable energy economy, from assembling turbines on the factory line to building and operating them in the field. It's about making America more energy-independent in an increasingly chaotic world. And it's about bringing much-needed stability to our unsettled global climate. With strong collaboration between government and the private sector, wind can truly become the heartland's new harvest.

Tables

Table 1. U.S. Onshore Wind Power Capacity by State in Gigawatts (GW) and Gigawatt Hours (GWH)

STATE	WIND ENERGY POTENTIAL[1]			CURRENT CAPACITY[2]	
	CAPACITY (GW)	RANK	OUTPUT/YR (GWH)	CAPACITY (GW)	RANK
Texas	1,901.5	1	6,527,850	10.085	1
Iowa	570. 7	7	2,026,340	3.675	2
California	34.1	20	105, 646	3.177	3
Minnesota	489.3	12	1,679,480	2.192	4
Washington	18.5	22	55,550	2.104	5
Oregon	27.1	21	80,855	2.104	6
Illinois	249.9	15	763,529	2.046	7
Oklahoma	516.8	9	1,788,910	1.482	8
North Dakota	770.2	6	2,983,750	1.424	9
Wyoming	552.1	8	1,944,340	1.412	10
Indiana	148.2	16	443,912	1.339	11
Colorado	387.2	13	1,288,490	1.299	12
New York	25.8	20	74,695	1.275	13
Kansas	952.4	2	3,646,590	1.074	14
Pennsylvania	3.3	29	9,673	.748	15
South Dakota	882.4	5	3,411,690	.709	16
New Mexico	492.1	11	1,644,970	.700	17
Wisconsin	103.8	17	300,136	.469	18
Missouri	274.4	14	810,619	.457	19

West Virginia	1.9	33	5,820	.431	20
Montana	944.0	3	3,228,620	.386	21
Idaho	18.1	23	52,118	.353	22
Maine	11.3	25	33,779	.266	23
Utah	13.1	24	37,104	.223	24
Nebraska	918.0	4	3,540,370	.213	25
Michigan	59.0	18	169,221	.164	26
Arizona	10.9	26	30,616	.128	27
Maryland	1.5	35	4,269	.070	28
Hawaii	3.3	29	12,363	.064	29
New Hampshire	2.1	34	6,706	.026	31[3]
Massachusetts	1.0	36	3,323	.018	32
Ohio	54.9	19	151,881	.011	33
Alaska	494.7	10	1,620,792	.010	34
Vermont	2.9	31	9,163	.006	36[4]
Arkansas	9.2	27	26,906	0	—
Nevada	7.2	28	20,823	0	—
Virginia	1.8	32	5,395	0	—
Other	2.2	—	6,411	.041	—
TOTAL	10,956.9	—	38,552,706	40.180	—

1. NREL, AWS Truewind, *80-Meter Wind Maps and Wind Resource Potential*, February 4, 2010, http://www.windpoweringamerica.gov/. Potential capacity is based on land areas where winds at 80 meters+ above ground would allow turbines to operate at a 30%+ capacity factor. States with less than 1 gigawatt of potential capacity are listed as "Other."
2. *AWEA U.S. Wind Industry Annual Market Report*—Year Ending 2010, figure 16, http://www.awea.org/.
3. Tennessee is ranked 30 in 2010, with 29 megawatts of installed capacity. It is listed as "Other" because its wind energy potential is less than 1 gigawatt (see note 1).
4. New Jersey is ranked 35 in 2010, with 8 megawatts of installed capacity. It is listed as "Other" because its onshore wind energy potential is less than 1 gigawatt. It should be noted, however, that New Jersey has substantial offshore wind energy potential.

Table 2. Top 10 Countries by Cumulative Installed Wind Power Capacity, December 31, 2010[1]

COUNTRY	MEGAWATTS	%
China	44,715	22.7
United States	40,180	20.4
Germany	27,214	13.8
Spain	20,676	10.5
India	13,065	6.6
Italy	5,797	2.9
France	5,660	2.9
United Kingdom	5,204	2.7
Canada	4,009	2.1
Denmark	3,752	1.9
Rest of the World	26,546	13.5
TOP 10 TOTAL	170,272	86.5
WORLD TOTAL	196,818	100

1. GWEC, *Global Wind Statistics 2010*, February 2, 2011, modified by updated reportage on China provided by CWEA, *China Wind Power Installed Capacity Data*, March 18, 2011.

Table 3. Top 10 Countries by Wind Power Capacity Installed during 2010[1]

COUNTRY	MEGAWATTS	%
China	18,928	49.5
United States	5,115	13.4
India	2,139	5.6
Spain	1,516	4.0
Germany	1,493	3.9
France	1,086	2.8

United Kingdom	962	2.5
Italy	948	2.5
Canada	690	1.8
Sweden	603	1.6
Rest of the World	4,750	12.4
TOP 10 TOTAL	33,480	87.4
WORLD TOTAL	38,230	100

1. GWEC, *Global Wind Statistics 2010*, February 2, 2011, modified by updated reportage on China provided by CWEA, *China Wind Power Installed Capacity Data*, March 18, 2011.

Table 4. U.S. Wind Power Installations by Manufacturer[1]

	ANNUAL INSTALLATIONS (IN MEGAWATTS)						% MARKET SHARE
MANUFACTURER	2005	2006	2007	2008	2009	2010	2009/2010
GE Wind (US)	1,433	1,146	2,342	3,585	3,995	2,543	40.0/49.7
Siemens (GER)	0	573	863	791	1,162	828	11.6/16.2
Gamesa (SP)	50	74	494	616	600	562	6.0/11.0
Mitsubishi (JP)	190	128	356	516	814	350	8.1/6.8
Suzlon (IN)	25	92	197	736	702	312	7.0/6.1
Vestas (DK)	700	439	948	1,120	1,490	221	11.2/4.3
Acciona (SP)	0	0	00	410	204	99	2.0/1.9
Clipper (US)	3	0	48	470	605	70	6.1/1.4
REpower (GER)	0	0	0	94	330	68	3.3/1.3
DeWind (GER)	NA	NA	NA	NA	NA	20	NA/0.4
Other	2	2	3	12	94	42	0.9/0.8
TOTAL	2,402	2,454	5,249	8,350	9,994	5,115	100.0

1. DOE, 2009 *Wind Technologies Market Report; AWEA U.S. Wind Industry Annual Market Report—Year Ending 2010.*

Table 5. Global Installed Wind Capacity (in Megawatts), 1996–2010[1]

YEAR	ANNUAL	CUMULATIVE
1996	1,280	6,100
1997	1,530	7,600
1998	2,520	10,200
1999	3,440	13,600
2000	3,760	17,400
2001	6,500	23,900
2002	7,270	31,100
2003	8,133	39,431
2004	8,207	47,620
2005	11,531	59,091
2006	15,245	74,052
2007	19,866	93,820
2008	26,560	120,291
2009	38,610	158,738
2010	38,230	196,818

1. GWEC, *Global Wind Statistics 2010,* modified by CWEA, *China Wind Power Installed Capacity Data,* March 18, 2011.

Acknowledgments

PEOPLE SAY THAT IT TAKES A VILLAGE to raise a child. Writing a book is not altogether different. So many people have helped at every stage—ranchers and farmers who warmly welcomed me onto their lands and into their homes, construction workers and factory laborers who let me take a close look at what they do, wind industry leaders and transmission experts who patiently tutored me in their trades, colleagues and friends in the environmental community who responded to my barrage of queries, a superb crew of research assistants, inspiring editorial guides, and a publishing team that has been masterful in every aspect of creating this book.

My agent, Colleen Mohyde at the Doe Coover Agency, reassuringly embraced my proposal and was wonderfully sure-footed in leading me to Beacon Press, a publisher that is dedicated to both the authors they take on and the ideas they are advancing. Alexis Rizzuto, my editor at Beacon, has been brilliant in keeping me focused on what readers really need to know about a subject that gets dangerously technical very fast. She and the book's entire Beacon team (Tom Hallock, Susan Lumenello, Reshma Melwani, Will Myers, and P. J. Tierney) are superb at what they do.

The pages of this book bear the imprints of several other creative forces. Larry Tye has been the best of role models, contagious in his love of book writing and a supportive friend throughout. Rosemary Ahern has been a gifted writing coach, helping a circumspect lawyer bring his own voice to the fore. My sister, Sally Bliumis-Dunn, brought a poet's eye and ear to her reading of the manuscript. Joel Segel aided my early search for a way to infuse a technical subject with life and humor.

From his perspective as a wind developer and lifelong environmentalist, Rob Gardiner combed every page for balance and technical accuracy. Seasoned journalist Ira Chinoy was equally vigilant in making sure my writing was well sourced. (I hope he'll forgive the rumored Napoleon quote that opens chapter 4. I liked it too much to delete it.) Dana Peck, a savvy wind project manager and long-ago coworker on Capitol Hill, scrutinized my writing on transmission issues, as did Seth Kaplan, a recent colleague at the Conservation Law Foundation. Renewable energy analyst Jan Hamrin filled out my understanding of California's wind energy boom and bust of the 1980s. Eric Lantz, Michael Milligan, and Suzanne Tegen at the National Renewable Energy Laboratory helped me grasp the economics of wind, as did Andrew Mills and Ryan Wiser at Lawrence Berkeley National Laboratory and Matt Kaplan of IHS Emerging Energy Research.

Researching this book took me to remote reaches of America and even more distant corners of the world. Horizon Wind Energy's Tanuj TJ Deora—now director of the Colorado Governor's Energy Office—accompanied my first visit to Meridian Way, where Michelle Graham educated me about the wind farm and was amazingly resourceful in introducing me to Cloud County landowners, educators, and civic leaders. Horizon CEO Gabriel Alonso and company staff across several states were instrumental in demonstrating what it takes to finance, build, and operate a wind farm.

Among other wind developers, Invenergy's Mark Leaman, together with Art Fletcher, arranged for me to spend a fascinating few days meeting company staff and neighbors at the Grand Ridge Wind Farm. Don Furman at Iberdrola Renewables was enlightening about everything from his company's wildlife protection innovations to the challenges of getting wind power onto the grid.

In the manufacturing realm, the Vestas team in Denmark and China showed me extraordinary generosity. Michael Holm lined up informative factory visits in Denmark and responded unfailingly to my multiple follow-up queries. I'm sure he's as relieved as I am that this book is finally going to press! In China, Andrew Hilton and Tu Trinh Thai made it possible for me to hear from Danish and Chinese managers about the challenges of fusing Western and Eastern workplace norms.

Mary Paul Jesperson and Pernille Florin Elbech at the Danish Embassy shaped my Denmark visit, together with Christian Wedekinck Olesen at the Climate Consortium. Valuable insights on wind energy's Danish ascent were offered by Parliament members Anne Grete Holmsgaard and Per Ørum Jørgensen, Lars Aagaard at the Danish Energy Association, and Rune Birk Nielsen at the Danish Wind Industry Association. I also want to thank Steen Hartvig Jacobsen for sharing his deep historical knowledge about Danish energy policies.

Armond Cohen at the Clean Air Task Force guided me wisely in preparing my trip to China, and the Natural Resources Defense Council's China staff provided useful background on Chinese wind manufacturing. Lin Wei and Ming Sung arranged informative meetings, capably interpreted by Houji Zhuzhu. Sebastian Meyer of Azure International shared keen observations and timely data on Chinese wind energy development. Along with all I learned from the entrepreneurs and analysts who appear or are cited in chapter 4, I am grateful for the insights offered by Alfred Zhao at the Chinese Wind Energy Association, Ryan Chen at the China New Energy Chamber of Commerce, Ellen Carberry at the China Greentech Initiative, and Jasmine Zhang at PriceWaterhouseCoopers.

Stateside, Bob Loyd at Clipper Wind was a blessing. He opened the door to some great discussions with his multigenerational crew at the Cedar Rapids assembly plant, and he was a rich resource on the city's long struggle to hold onto a viable industrial base. Bob Gates, in Clipper's California headquarters, brought decades of wind energy experience to his reflections on the company's technology innovations, its financial hurdles, and the damage caused by erratic federal policies.

Other manufacturers who helped me understand the wind trade were Jim Charmley, Hans Landin, and Lorrie Paul Crum at Timken; Daniel McGahn and Jason Fredette at American Superconductor; Bob Paxton at Broadwind; John Grabner at Cardinal Fastener; and Dheeraj Choudhary at Parker Hannifin. Richard Stuebi at the Cleveland Foundation kindly steered me toward several of these fine people.

General Electric eluded my persistent efforts over several months to arrange factory visits and interviews with company officials. This was unfortunate, given GE's central role in supplying America's turbines. Nevertheless, Millissa Rocker did provide useful speeches and testimony by Vic Abate, vice president of GE Energy's renewables division.

A successful wind industry doesn't just depend on the companies that produce the technology and put up the turbines. I was privileged to meet several economic development boosters who have catalyzed factories and wind farms in their communities: Mayor Charles "Chaz" Allen of Newton, Iowa; Joe Jongewaard and Beth Govoni at the Iowa Department of Economic Development; Dennis Jordan at the Cedar Rapids Chamber of Commerce; CloudCorp's Kirk Lowell in Concordia, Kansas; Connie Neininger at the White County Economic Development Organization in Indiana; and Matt Sorensen at the McLean County Commission in Illinois. I also was inspired by the educators who are racing to create a well-trained corps of wind technicians: Bruce Graham at Cloud County Community College, Ahmad Hemami at Iowa Lakes Community College, and P. Barry Butler at the University of Iowa's College of Engineering top the list.

In exploring the energy industry's wildlife impacts, I was aided by several key people beyond those mentioned in chapter 7. Michelle Carder and Matt Gasner of Western Ecosystems Technology took me into the field to see how they monitor bats at Indiana's Meadow Lake Wind Farm. Shannon Anderson at the Powder River Basin Resource Council and the Sierra Club's Brad Mohrmann filled me in on the ecological devastation wrought by Wyoming's coal, gas, oil, and uranium industries. John Emmerich, executive director of the Wyoming Game and Fish Department, described his agency's efforts to protect the sage grouse and other wildlife.

Local battles over wind farm siting took me to the Flint Hills of Kansas, the Northern Laramie Range in Wyoming, and Vermont's Taconic Mountains. The Flint Hills controversy is well covered in the text, but I am especially grateful to Kansas Audubon's Ron Klataske for introducing me to local ranchers and conservation experts. I also thank ranchers Rose Bacon and Pete Ferrell for sharing their very different perspectives on a hotly divisive issue. John Briggs of the Konza Prairie Biological Station, Rob Manes and Brian Obermeyer of the Nature Conservancy, and Chuck Rice of Kansas State University gave me valued tutorials on tallgrass prairie ecology. Broader insights on wind energy's importance to Kansas came from Scott Allegrucci of the Great Plains Alliance for Clean Energy; Steve Baccus and Mike Irvin at the Kansas Farm Bureau; Rod Bremby, former secretary of the Kansas Department of Health and Environment; and Nancy Jackson at the Climate and Energy Project.

Securing wind energy's place in a state as richly endowed with fossil and nuclear fuels as Wyoming is no small undertaking. In probing that state's policies, I benefited hugely from my conversations with state senate president Jim Anderson; Cheryl Riley and David Picard of the Wyoming Power Producers Coalition; and Aaron Clark, who advised former governor Dave Freudenthal on energy policy. Although the hot debate over wind farm siting in the Northern Laramies didn't find its way into this book, I learned much about wind energy politics from rancher Rick Grant, Ken Lay of the Northern Laramie Range Alliance, and Ed Werner, chair of the Converse County Commission until wind farm opponents voted him out of office. I also spent an eye-opening day with L.J. and Karen Turner, Campbell County ranchers whose cattle graze on the edge of America's largest open-pit coal mine. I wish there was enough room in this book to tell every amazing story I encountered.

In Vermont's Taconic Mountains, concerns about noise and viewshed protection have dominated the debate about wind. Ken Kaliski of the Resource Systems Group was hugely instructive on the science of acoustics, and Dr. Michael Nissenbaum was forthcoming in explaining his own research on noise issues. I am also indebted to Annette Smith of Vermonters for a Clean Environment for her tour of the contested Vermont Community Wind site west of Rutland.

Power transmission is a hugely technical, bureaucratically intertwined realm that several people made it easier to fathom: Brad Beecher of Empire District Electric; Jay Caspary and Carl Monroe of the Southwest Power Pool; Craig Cox at the Interwest Energy Alliance; Loyd Drain at the Wyoming Infrastructure Authority; Steve Gaw of the Wind Coalition; Mark Fagan of the Harvard Kennedy School; Chris Miller and Bri West at the Piedmont Environmental Council; John Nielsen of Western Resources Advocates; Todd Parfitt and Tom Schroeder of Wyoming's Department of Environmental Quality; Chris Petrie at the Wyoming Public Service Commission; and David Smith at TransWest Express LLC. Jon Wellinghoff, chairman of the Federal Energy Regulatory Commission, brought his powerful vision to our meeting, and Julia Bovey proved herself a true friend in arranging that interview.

The obstacles to getting truly transformative energy policies in place came to light in discussions with U.S. representatives Ed Markey and Henry Waxman. These two mavericks have fought long and hard to give our global climate crisis the priority it deserves. Hopefully, their day will come. David Osterberg of the Iowa Public Policy Project was uplifting in describing what his own state has done to bring renewable energy into the mainstream.

In addition to holding the most informative conferences and workshops in the trade, the American Wind Energy Association has a bright staff of professionals who have been highly responsive to my research needs. My particular thanks go to Rob Gramlich, Kathy Belyeu, Hans Detweiler, Michael Goggin, and Elizabeth Salerno.

Each of my trips yielded new questions and a new crop of interviews to be transcribed. I was lucky to have a spirited and utterly reliable group of research assistants at the ready. Kate Davies uncovered a trove of studies on bird and bat concerns, turned recordings of varying quality into precise transcripts, and offered astute insights on the evolving manuscript. My thanks also go to Kathryn Boucher, Christine Cho, Michael Dorsi, Ari Peskoe, Christina Putz, and Allie Rosene-Mirvis. Vida Margaitis and Diane Sredl, librarians at Harvard University, were ever-resourceful in tracking down valuable government documents. I am grateful as well to my mother, Joan Warburg, for her many phone calls and e-mails alerting me to wind energy coverage in the media.

My wife, Tamar, has been both a tireless editor and an endless source of support and encouragement. Reading numerous drafts of this book, she praised generously, but she also dared to deliver the difficult news of a chapter's need for fundamental reshaping. I'm sure that she and our daughters, Tali and Maya, have heard more about wind energy than they ever wanted to know.

Speaking of Tali and Maya, this book is dedicated to them. While plenty of work still lies ahead for my generation, theirs will have to carry on the tough work of steering our national and global energy choices onto a saner course. I only hope that we're all up to the challenge.

Notes

INTRODUCTION

1. Letter from Franklin D. Roosevelt to James P. Warburg, May 23, 1934, quoted in James P. Warburg, *The Long Road Home* (New York: Doubleday & Co., 1964), 156–57.

CHAPTER ONE: CLOUD COUNTY REVIVAL

1. See table 1, estimating total yearly wind generation potential in Kansas at 3.6 million gigawatt hours. U.S. power generation in 2010 totaled 4.1 million gigawatt-hours. U.S. Energy Information Administration (EIA), table 1.1, "Net Generation by Energy Source: Total (All Sectors)," released March 11, 2011, http://www.eia.gov/.

2. American Wind Energy Association, *U.S. Wind Industry Annual Market Report,* 12, fig. 12.

3. T. Lindsay Baker, *Field Guide to American Windmills,* 33–40, 339–421.

4. Kansas Statutes 79–201 (effective January 1, 1999). The federal production tax credit, more fully described in chapter 3, "Rust Belt Renewables," had been intermittently available since passage of the Energy Policy Act of 1992 (P.L. 102–486).

5. Duane Schrag, "Wind Turbines Could Resume Spinning Soon," *Salina (KS) Journal,* April 28, 2009, http://www.saljournal.com/, citing EIA data.

6. Kurt and Helen Kocher, interviews with author, March 19, 2009, and May 14, 2009.

7. Ray Mason, interview with author, May 13, 2009.

8. Monica Perin, "Wind Energy Visionary Flies under the Radar," *Houston Business Journal,* April 27, 2003, http://www.bizjournals.com/.

9. Jim Hoy, "Environmental Price Tag Too High for Wind Farms," undated, Protect the Flint Hills, www.protecttheflinthills.org/, accessed September 30, 2011.

10. Tim Unruh, "Company Finds One of the Answers to the Energy Crisis Is Blowing in the Wind," *Salina Journal,* October 8, 2008, http://www.salina.com/.

11. Tom Fowler, "Zilkha Wind Firm Purchased by Goldman Sachs," *Houston Chronicle,* March 22, 2005, http://www.chron.com/.

12. Sergio Goncalves, "EDP to Buy $2.2 bln U.S. Horizon Wind Energy," Reuters, March 27, 2007, http://www.reuters.com.

13. Jim Roberts, senior project manager, Horizon Wind Energy, phone interviews with author, May 12, 2009, and June 17, 2009.

14. See Vestas, *V90–3.0 MW: An Efficient Way to More Power,* http://www.autonavz duch.cz/, accessed April 4, 2011.

15. Carole Engelder, director of construction management, Midwest Region, Horizon Wind Energy, phone interview with author, June 11, 2009.

16. Roberts phone interview, May 12, 2009. Although he did not discuss per-turbine payments in our conversations, Roberts is cited in media reportage as giving $4,000 to $8,000 per turbine as the range of payments made to Meridian Way landowners. Tim Unruh, "Company Finds One of the Answers to the Energy Crisis Is Blowing in the Wind," *Salina Journal*, October 8, 2008, http://www.salina.com/.

17. Bonnie Sporer, Cloud County landowner, phone interview with author, May 18, 2009.

18. Carole Engelder, e-mail to author, June 19, 2009. Roughly 47 percent of the workforce was hired locally.

19. Kirk Lowell, executive director, Cloud County Development Corporation, interview with author, May 20, 2009.

20. Bruce Graham, chairman, Wind Energy Department, Cloud County Community College, interview with author, March 19, 2009, and e-mail to author, March 4, 2011.

CHAPTER TWO: EARLY ADOPTERS

1. Danish Commission on Climate Change Policy, *Green Energy*, 32.

2. BTM Consult, *World Market Update 2004 (Forecast 2005–2009)*, March 31, 2005, http://www.btm.dk/.

3. John Acher, "UPDATE 2—China Rivals Narrow Gap on Wind Leader Vestas," Reuters, March 15, 2011, http://af.reuters.com/.

4. Vestas, *Annual Report—2010*, http://www.vestas.com/. Conversion to U.S. dollars at exchange rate applicable on December 31, 2010.

5. Lone Mortensen, director of people and culture, Blades Division, Vestas Wind Systems A/S, Lem, Denmark, October 7, 2009.

6. See Vestas annual reports, 2008–2010, http://www.vestas.com/.

7. Danish Energy Authority, "Combined Heat and Power Production in Denmark," undated fact sheet, http://www.ambottawa.um.dk/, accessed August 1, 2011.

8. Buen, "Danish and Norwegian Wind Industry," 3890.

9. Wind power producers were paid 85 percent of the pretax price charged to consumers. Birger T. Madsen, "Public Initiatives and Industrial Development after 1979," in Nissen, *Wind Power*, 52, 54.

10. Toke, "Wind Power in UK and Denmark," 83 et seq.

11. Van Est, *Winds of Change*, 89. This rule was issued in 1985.

12. Buen, "Danish and Norwegian Wind Industry," 3890.

13. Sijm, *Performance of Feed-in Tariffs*, 11.

14. Gipe, "Wind Energy Comes of Age," 760.

15. "Denmark Struggles with Regulations—So Far, So Bad," *Windpower Monthly*, April 1, 2000, http://www.windpowermonthly.com/. See also Szarka, "Wind Power," 3041, 3046.

16. Toke, "Wind Power in UK and Denmark," 92.

17. Van Est, *Winds of Change*, 87.

18. Richard M. Nixon, Special Message to the Congress on the Energy Crisis, January 23, 1974, http://www.presidency.ucsb.edu/.

19. Richard M. Nixon, State of the Union address, January 30, 1974, http://abcnews.go.com/.

20. U.S. Energy Information Administration (EIA), "Petroleum & Other Liquids: U.S. Imports by Country of Origin," released July 27, 2010, http://www.eia.doe.gov/.

21. Jimmy Carter, Report to the American People on Energy, February 2, 1977 http://millercenter.org/.

22. Jimmy Carter, Address to the Nation on Energy, April 18, 1977, http://millercenter.org/.

23. Denis Hayes, phone interview with author, July 14, 2010. See also D. Hayes, *Rays of Hope*, 174–80.

24. PURPA, sec. 210. The Federal Energy Regulatory Commission (FERC) encouraged states to set rates at the full "avoided cost" of generating the same increment of power in the absence of the independent power sources. See Gipe, *Wind Energy Comes of Age*, 31.

25. See Thomas A. Starrs, "Legislative Incentives and Energy Technologies," 129–34, describing FERC v. Mississippi, 456 U.S. 742 (1982), and American Paper Institute, Inc. v. American Electric Power Service Corp., 461 U.S. 402 (1983).

26. Energy Tax Act of 1978, PL 95–618, 92 STAT. 3195, Sec. 301(a). See also Minan, "Encouraging Solar Energy Development," 31–34.

27. Crude Oil Windfall Profits Tax of 1980, PL 96–223, 94 STAT. 230, Sec. 221(a).

28. Fred Branfman, "Moving toward the Abyss," *Salon*, April 18, 1996, http://www.salon.com/.

29. Lovins, *Soft Energy Paths*, 55.

30. See Clark, *Energy for Survival*.

31. See U.S. House of Representatives, *Renewable Energy Incentives*, 147 (testimony of Warren D. Noteware, California Energy Commission).

32. Righter, *Wind Energy in America*, 208.

33. Ellen Paris, "The Great Windmill Tax Dodge," *Forbes*, March 12, 1984, 40.

34. Gipe, *Wind Energy Comes of Age*, 31.

35. See Righter, *Wind Energy in America*, 215; Starrs, "Legislative Incentives and Energy Technologies," 109, table 1; Smith, "Wind Farms of the Altamont Pass Area," 153, fig. 3.

36. Righter, *Wind Energy in America*, 158. For testimony by representatives of NASA and major aerospace firms, see U.S. House of Representatives, *Wind Energy Systems Act of 1980*.

37. Gipe, *Wind Energy Comes of Age*, 106.

38. Ibid., 96.

39. See Asmus, *Reaping the Wind*, 87–131.

40. Van Est, *Winds of Change*, 54.

41. Righter, *Wind Energy in America*, 183, quoting interview with Paul Gipe, November 11, 1992.

42. Danish Wind Industry Association, *Denmark—Wind Power Hub: Profile of the Danish Wind Industry*, 2008, 13, http://www.windpower.org/.

43. Van Est, *Winds of Change*, 59; Righter, *Wind Energy in America*, 181.

44. Powerplant and Industrial Fuel Use Act of 1978, PL 95–620, 42 USC 8301.

45. Powerplant and Industrial Fuel Use Act Amendments of 1987, PL 100–42, 42 USC 8312.

46. See Gipe, "Wind Energy Comes of Age," 738; Starrs, "Legislative Incentives and Energy Technologies," 119.

47. Ronald Reagan, "Acceptance Speech at the 1980 Republican Convention," July 17, 1980, http://www.nationalcenter.org/.

48. Sissine, *Renewable Energy: A New National Commitment?*, unnumbered table, "Dept. of Energy (DOE) Renewable Energy R&D Funding: FY74 to FY92."

49. Gipe, "Wind Energy Comes of Age," 758, fig. 4.

50. Almost 70 percent of the world's wind-generated electricity came from California's big-three wind farms in 1985. Smith, "Wind Farms of the Altamont Pass Area," 146.

51. Van Est, *Winds of Change*, 92.

52. See European Monitoring Centre on Change, *EMCC Case Studies: Energy Sector: Vestas, Denmark*, European Foundation for the Improvement of Living and Working Conditions, 2008, 4, http://www.eurofound.europa.eu/; and *Reference for Business, Encyclopedia for Business*, 2d ed., "Vestas Wind Systems A/S," http://www.referenceforbusiness.com/.

53. See Sijm, *Performance of Feed-in Tariffs*, 7–8 and table 3.2.

54. Ibid., table 3.2; DOE, U.S. Installed Wind Capacity, 2000 Year End Wind Power Capacity (map), http://www.windpoweringamerica.gov/.

55. Peter Wenzel Kruse, senior vice president for group communication and investor relations, Vestas, interview with author, October 8, 2009.

56. With 327 megawatts of new wind power capacity in 2010, Denmark ranked eleventh among European nations that year. European Wind Energy Association, *Wind in Power: 2010 European Statistics*, February 2011, http://www.ewea.org/.

57. Ditlev Engel, press conference on Q3 accounts 2010, Radisson Blue Royal Hotel, Copenhagen, October 26, 2011. Additional information provided via e-mail to author by Michael Holm, public relations manager, Vestas, October 26, 2010, and June 10, 2011.

58. See American Wind Energy Association, *U.S. Wind Industry Annual Market Report*, 36.

CHAPTER THREE: RUST BELT RENEWABLES

1. "President-Elect Obama Visits and Speaks at Cardinal Fastener, Bedford Heights, OH," January 16, 2009, YouTube, http://www.youtube.com/.

2. American Wind Energy Association (AWEA), *U.S. Wind Industry Annual Market Report*, 37.

3. Beth Govoni and Elizabeth Conley, Iowa Department of Economic Development, e-mail to author, August 4, 2010.

4. "Stimulus Money IS Creating US Jobs: American Wind Energy Association Response to American University Study/ABC World News Story," *Electric Energy Online*, February 12, 2010, http://www.electricenergyonline.com.

5. Elizabeth Salerno and Jessica Isaacs, "The Economic Reach of a Thriving Sector," *Windpower Monthly, Special Report–United States: Wind Industry a Driver of Economic Recovery*, May 2009, 7.

6. Interviews with Bob Loyd, plant manager, and employees at Clipper Wind Turbine Works were conducted during the author's site visit to Cedar Rapids, IA, February 2, 2010.

7. David Pitt, "Historic Maytag Factory Shuts Its Doors," Associated Press, October 25, 2007, http://www.manufacturing.net/.

8. Robert Gates, chief commercial officer, Clipper Windpower, phone interview with author, April 13, 2010. See also Jesse Broehl, "Domestic Supply Chain Must Grow," *Windpower Monthly, Special Report–United States: Wind Industry a Driver of Economic Recovery*, May 2009, 11.

9. "Drive Train Retrofit Means More Delays—Further Bad News from Clipper on Liberty Component Problems," *Windpower Monthly*, October 2007, 42; "Series of Blade Repairs Under Way—More Clipper Problems," *Windpower Monthly*, February 2008, 39; "Growing Revenue but Higher Losses," *Windpower Monthly*, June 2009, 31.

10. "Growing Revenue but Higher Losses," *Windpower Monthly*, June 2009, 31; "Analysis—Changing Blades without a Tall Crane," *Windpower Monthly*, April 2010, 22.

11. See Karl-Erik Stromsta, "Pertz Quits Over Losses," *Recharge News*, March 12, 2010, 10; Robin Pagnamenta, "Clipper Has the Wind beneath Its Wings after US Funding Deal," *Times (London)*, December 11, 2009, http://business.timesonline.co.uk; Ben Backwell, "Clipper in a Cash Crisis," *Recharge*, September 24, 2010, www.rechargenews.com/; and Associated Press, "United Technologies Optimistic about Wind Power," February 10, 2011, http://www.bloomberg.com/.

12. "Clipper Teams Up to Forge New Beginning," *Windpower Monthly*, May 2010, 77.

13. Gates phone interview, April 13, 2010.

14. The production tax credit, under section 1914 of the Energy Policy Act of 1992 (P.L. 102–486), originally targeted wind and biomass derived from field crops or trees grown exclusively for power production. Other forms of renewable energy, including solar, geothermal, landfill gas, trash combustion, and certain kinds of small-scale hydropower, were added later. Sissine, *Renewable Energy: Tax Credit, Budget, and Electricity Production Issues*, 4.

15. The production tax credit for wind has since been raised to 2.2 cents per kilowatt hour. "Federal Incentives/Policies for Renewables & Efficiency: Renewable Electricity Production Tax Credit," Database of State Incentives for Renewables and Efficiency (DSIRE), rev. April 25, 2011, http://www.dsireusa.org/.

16. See H.R. 1, P.L. 111–5, Sec. 1101. The PTC was raised again in 2010 to 2.2 cents per kilowatt hour. For further analysis, see Viva Hammer, "Special Report: Alternative Energy Gets a Second Chance," *Tax Notes*, November 22, 2010, http://vivahammertax.com/.

17. See Dan Eggen, "Four Democratic Senators Aim to Halt Stimulus Wind Project," *Washington Post*, March 4, 2010.

18. AWEA, *U.S. Wind Industry Annual Market Report*, 36; *AWEA Year-End 2009 Market Report*, 37.

19. Under the Price-Anderson Act, each licensed reactor pays for $375 million in private insurance covering off-site liability. If an accident were to cause damage above that amount, all 104 licensed U.S. reactors would be assessed a pro rata share of the excess, up to a maximum of $111.9 million apiece. Beyond the $11.6 billion covered by this industrywide pool, state and local governments would have to petition Congress for supplemental disaster relief. See Nuclear Regulatory Commission, "Fact Sheet on Nuclear Insurance and Disaster Relief Funds," June 2011, http://www.nrc.gov/.

20. For a general overview of this company, see Pruitt, *Timken*.

21. Lorrie Paul Crum, communications department, and James E. Charmley, director of product technology, Timken, interviews with author, March 5, 2010.

22. Ward J. "Tim" Timken Jr., "Private Enterprise: A Cornerstone of American Democracy," Major Issues Lecture, Ashland University, Ashbrook Center for Public Affairs, October 16, 2009, http://www.ashbrook.org/.

23. H. Josef Hebert, "Timken, Belden, and Their Workers Keep an Eye on Climate Bills," Associated Press, October 11, 2009, http://www.cantonrep.com/.

24. Crum, e-mail to author, December 10, 2009.

25. Associated Press, "GE Partners to Create Wind Farm in Lake Erie Near Cleveland; Aims for 1,000 MW Farm by 2020," May 24, 2010.

CHAPTER FOUR: THE CHINESE ARE COMING

1. The top fifteen global turbine manufacturers in 2010, by rank and market share, were Vestas (Denmark) #1–12%; Sinovel (China) #2–11%; GE (U.S.) #3–10%; Goldwind (China) #4–10%; Enercon (Germany) #5–7%; Gamesa (Spain) #6–7%; Dongfang (China) #8–7%; Suzlon (India) #9–6%; Siemens (Germany) #10–5%; United Power (China) #10–4%; Mingyang (China) #11–3%; REpower (Germany) #12–2%; Sewind (China) #13–2%; Nordex (Germany) #14–2%; XEMC (China) #15–1%. Reuters, "UPDATE 2-China Rivals Narrow Gap on Wind Leader Vestas," March 15, 2011, http://www.reuters.com/, citing data from MAKE Consulting.

2. This conference, attended by the author, was held on October 21–23, 2009.

3. This ceremony, which took place on October 15, 2009, featured remarks by Lars Anderson, president, Vestas-China; Ni Xiangyu, vice chairman, the Tianjin Economic-Technological Development Area; and others.

4. Chinese Wind Energy Association (CWEA), *China Wind Power Installed Capacity Data (2010)*, March 18, 2011 (in Chinese), table 2; Global Wind Energy Council, *Global Wind Statistics 2010*, http://www.gwec.net/.

5. China's total electricity consumption is predicted to reach 8,200 terawatt-hours by 2020, with total installed capacity growing to 1.885 terawatts. "China Electricity Consumption to Almost Double by 2020: China Electricity Council," Xinhua News Agency, December 21, 2010, http://news.xinhuanet.com/.

6. McElroy et al., "Potential for Wind-Generated Electricity in China," 1378–80.

7. See Keith Bradsher, "China Outpaces U.S. in Cleaner Coal-Fired Plants," *New York Times*, May 11, 2009.

8. U.S. Energy Information Administration (EIA), "Independent Statistics and Analysis," July 2009, http://www.eia.doe.gov/.

9. See "China's Coal Demand Likely to Hit 3.8 bln t [billion metric tons] in 2015," June 9, 2010, iStockAnalyst, www.istockanalyst.com/. Projected coal use of 3.8 billion metric tons converts to 4.18 billion "short" U.S. tons.

10. The official death toll for Chinese coal miners was 3,215 in 2008 and 2,631 in 2009. "China Mine Accidents Multiply—28 Dead, 192 Missing," Agence France-Presse, April 2, 2010, http://www.chinamining.org/.

11. World Nuclear Association, "World Nuclear Power Reactors & Uranium Requirements," table, March 2, 2011, http://www.world-nuclear.org/.

12. See Antoaneta Bezlova, "'Three Gorges Dam May Displace Millions More," IPS-Inter Press Service, October 12, 2007, http://ipsnews.net/; Jim Yardley, "Chinese Dam Projects Criticized for Their Human Costs," *New York Times*, November 19, 2007.

13. Lema, "Between Fragmented Authoritarianism and Policy Coordination," 3882.

14. See Li, *Study on the Pricing Policy of Wind Power in China*, 7.

15. Renewable Energy Law of the People's Republic of China, approved by the Standing Committee of the National People's Congress, February 28, 2005, articles 6–9.

16. Wu Qi, "Landmark Project Heralds Grand Vision," *Windpower Monthly Special Report–China: Market Ambition Ramps Up a Gear*, October 2009, 20–21. See also Louis

Schwartz, "China's New Generation: Driving Domestic Development," March 10, 2009, RenewableEnergyWorld.com, http://www.renewableenergyworld.com/.

17. Yvonne Chan, "China Unveils $140bn Plan to Build Seven Giant Wind Farms by 2020," *BusinessGreen*, June 30, 2009, http://www.businessgreen.com/. See also Rujun Shen and Tom Miles, "China's Wind-Power Boom to Outpace Nuclear by 2020," Reuters, April 20, 2009, http://uk.reuters.com/.

18. Wu Qi, "New Milestone for Domestic Leader," *Windpower Monthly*, May 2009, 31.

19. Wilson Guo, China country manager, UPC, interview with author, October 20, 2009.

20. Charles R. McElwee II, Squire Sanders & Dempsey LLP, phone interview with author, November 23, 2009.

21. See Lewis, *Comparison of Wind Power Industry Development Strategies*, 15–17.

22. Goldwind Science & Technology Company, *Company Overview and Highlights*, September–October 2009, 12, www.goldwindglobal.com.

23. Chunhua Li, director, International Business Department, Goldwind, and Eva Xie, vice director, Investment Department, Tianrun, interview with author, October 18, 2009; Liang Xuan, International Business Department, Goldwind, e-mail to author, June 30, 2010.

24. See Wu Qi, "Grid Quotas Aim to Connect 9 GW of Stuck Chinese Wind," *Windpower Monthly*, June 2010, 17; Li Jing, "China Plans for Renewable Energy," *China Daily*, August 25, 2009, http://www.chinadaily.com.cn/.

25. "China Electricity Consumption to Almost Double by 2020," *China Daily*, December 22, 2010.

26. Lewis, *A Review of the Potential International Trade Implications*, 3.

27. Azure International, Beijing, data provided to author via e-mail, September 30, 2010.

28. CWEA, *China Wind Power Installed Capacity Data (2010)*, March 18, 2011 (in Chinese), table 2. Rankings are based on megawatts installed in 2010.

29. See Office of the U.S. Trade Representative, "United States Requests WTO Dispute Settlement Consultations on China's Subsidies for Wind Power Equipment Manufacturers," December 22, 2010, http://www.ustr.gov/. See also Keith Bradsher, "To Conquer Wind Power, China Writes the Rules," *New York Times*, December 14, 2010.

30. See Office of the U.S. Trade Representative, "United States Requests WTO Dispute Settlement Consultations."

31. Chunhua Li interview, October 15, 2009.

32. Wu Qi, "Goldwind Wins New US Orders," *Windpower Monthly*, April 25, 2011.

33. CWEA, *China Wind Power Installed Capacity Data (2010)*, table 2.

34. See Lyn Harrison and Eric Prideaux, "Conference Report China: World Asks China to Drop Its Barriers," *Windpower Monthly*, 57, 60.

35. Matthew Kaplan, associate director, North American Wind Energy Markets, IHS Emerging Energy Research, interview with author, June 7 2011.

36. Jim Bai and Chen Aizhu, "China Wind Power Capacity Could Reach 1,000 GW by 2050," Reuters, October 19, 2011, http://www.reuters.com/.

37. Friedman, *Hot, Flat, and Crowded*, 372–73.

CHAPTER FIVE: WORKING THE WIND

1. Ira Chinoy, associate professor, Philip Merrill College of Journalism, University of Maryland, e-mail to author, March 1, 2011.

2. *Inside Indiana Business*, "More Wind Turbine Components Arrive at Port of Indiana," June 19, 2009, http://www.insideindianabusiness.com/.

3. Jesse Broehl, "Transportation Is the Wind beneath U.S. Industry Wings," *Windpower Monthly Special Report: Freight on Board*, July 2010, 5.

4. "Western States—Four Ports Make US Top Ten List," *Windpower Monthly, Special Report: U.S. Investment and Development*, September 2010, 23.

5. "Texas and Oklahoma Lead Quiet Region," *Windpower Monthly, Special Report: U.S. Investment and Development*, September 2010, 16.

6. "Trains Deliver the Wind," *Breakbulk Industry News*, September 22, 2009; "Iowa Cornfield Sprouts Wind Towers," *Breakbulk Industry News*, September 22, 2009, http://www.breakbulk.com/ (archival access limited to subscribers).

7. Mark Anderson, "Trucking Turbines Gets Sophisticated," *Windpower Monthly*, April 2009, 92.

8. Site visit to Grand Ridge Wind Farm, Ransom, IL; Adam Hartman, construction manager, Invenergy, interview with author, May 8, 2009.

9. Alvin Cargill, Horizon Wind Energy, project manager for construction—Meadow Lake Wind Farm, White and Benton counties, IN, interview with author, July 22, 2009.

10. Bruce Graham, presentation on Meridian Way Wind Farm to Kansas–Nebraska Radio Club, Cloud County Community College, observed by author, March 19, 2009.

11. American Wind Energy Association, "Supply Chain—Anatomy of a Wind Turbine," http://www.awea.org/.

12. See "President-Elect Obama Visits and Speaks at Cardinal Fastener, Bedford Heights, OH," January 16, 2009, YouTube, http://www.youtube.com/.

13. Clinton Newbold, quality assurance manager, Barnhart, at Meadow Lake Wind Farm, interview with author, July 23, 2009.

14. Martin Culik, Horizon Wind Energy, project manager for development, Meadow Lake Wind Farm, interview with author, July 23, 2009. See also Horizon Wind Energy website, Meadow Lake Wind Farm, http://www.horizonwind.com/.

15. Connie Neininger, economic development director, White County, IN, interview with author, July 18, 2009.

16. Steve Maples, Barnhart Crane & Rigging, site manager, Meadow Lake Wind Farm, interview with author, July 23, 2009. All subsequent quotes attributed to Steve Maples are from the same source.

17. Justin Van Beusekom, assistant operations manager, Meridian Way Wind Farm, Horizon Wind Energy, May 15, 2009.

18. Leo Jessen, lead technician, Grand Ridge Wind Farm, Invenergy, interview with author, May 10, 2009.

19. Ibid.

20. American Wind Energy Association, *AWEA Year-End 2009 Market Report*, 37–38.

21. See American Wind Energy Association, "AWEA Seal of Approval Program," http://archive.awea.org/, accessed April 11, 2011.

22. Cloud County Community College Grants List—Wind Energy Technology Training Program, provided to author by Bruce Graham, March 2010.

23. Brad Lowell, "CCCC Becoming Leader in Wind Energy Education," *Concordia (KS) Blade-Empire*, March 10, 2010, quoting Kim Krull, vice president for academic affairs, Cloud County Community College.

24. Acme Idea Company, New Duracell Smart Power Commercial, December 4, 2009, YouTube, http://www.youtube.com/.

25. P. Barry Butler, dean, College of Engineering, University of Iowa, interview with author, February 1, 2010.

26. Interviews with students, Wind Energy and Turbine Technology Program, Iowa Lakes Community College, conducted by author, February 5, 2010.

27. Michelle Graham, Horizon Wind Energy, operations administrator, Meridian Way Wind Farm, e-mail to author, May 16, July 9, December 17, 2009, and September 4, 2010.

28. Bruce Graham, e-mail to author, March 4, 2011.

29. Ahmad Hemami, instructor, Iowa Lakes Community College, interview with author, February 5, 2010.

30. Loma Roggenkamp, O&M technician, Siemens Energy, phone interview with author, September 7, 2010; e-mail to author, July 27, 2011.

31. Kristen Graf, executive director, Women of Wind Energy, phone interview with author, September 3, 2010.

32. AWEA Board of Directors, http://www.awea.org/, accessed April 11, 2011.

33. Jeanna Walters, former student, Wind Energy Technology Training Program, Cloud County Community College, phone interview with author, May 13, 2009.

34. U.S. Department of Energy, *Wind Power in America's Future*, 12.

35. Ibid., appendix C: "Wind-Related Jobs and Economic Impact," 207–10. Data are drawn, in part, from the explanatory text accompanying fig. C-7, with specific numerical projections underlying fig. C-7 provided by Suzanne Tegen, PhD, senior energy analyst, Strategic Energy Analysis Center, National Renewable Energy Laboratory (NREL), e-mail to author, August 31, 2010. Indirect jobs are based on "the increase in economic activity that occurs . . . when a contractor, vendor, or manufacturer receives payment for goods or services and in turn is able to pay others who support their business." Induced employment is caused by "the changes in wealth that result from spending by people directly and indirectly employed by the project." Ibid., 201.

CHAPTER SIX: THE PATH TO CLEANER ENERGY

1. Ryan H. Wiser, Lawrence Berkeley National Laboratory, "Tracking the US Wind Industry: Update on Cost, Performance, and Pricing Trends," AWEA Windpower 2011 Conference, May 23, 2011. In previously published government data, wind was credited with a more modest 2.3 percent of total U.S. electricity supply in 2010: 94,647 gigawatt hours out of 4,120,028 gigawatt hours. U.S. Energy Information Administration (EIA), *Electric Power Monthly*, table 1.1, accessed July 2011, http://www.eia.doe.gov/

2. EIA, *Electric Power Monthly*, table 1.1A. Solar-generated power totaled 1.3 gigawatt hours in 2011.

3. By way of comparison, a coal plant's capacity factor is typically in the 70 to 90 percent range, and a nuclear plant can operate at over 90 percent of its stated capacity.

4. National Renewable Energy Laboratory (NREL) and AWS Truewind, "Estimates of Windy Land Area and Wind Energy Potential by State for Areas >=30% Capacity Factor at 80m," February 4, 2010, U.S. Department of Energy (DOE), Wind Powering America, http://www.windpoweringamerica.gov/.

5. See David Appleyard, "Record Growth for EU Offshore Wind in 2010," *Renewable Energy World*, February 15, 2011, http://www.renewableenergyworld.com/. Europe added

883 megawatts in new offshore wind capacity in 2010, about 10 percent of total European wind installations that year.

6. Schwartz, *Assessment of Offshore Wind Energy Resources*, 3–4, table 1. This study understates the total offshore wind resource because it does not include data for Florida, Alabama, or Mississippi; wind maps for these states were unavailable. On the other hand, over 60 percent of the wind resources in NREL's offshore database are in waters deeper than 150 feet, where wind energy development is not yet technically feasible.

7. Vattenfall, *Life-Cycle Assessment*, 2.

8. Spath, *Life Cycle Assessment*, 10–13, 29, table 29.

9. Estimates of coal power plant CO2 emissions were calculated by David Schoengold, senior consultant, MSB Energy Associates, based on an analysis of DOE data for 2008. David Schoengold, e-mail to author, October 5, 2010; Armond Cohen, executive director, Clean Air Task Force, e-mail to author, September 30, 2010.

10. Vestas, *Lifecycle Assessment of a V90–3.0 MW Onshore Wind Turbine*, 7.

11. Vestas, *Lifecycle Assessment of Offshore and Onshore Sited Wind Power Plants*, 35–40.

12. DOE, *Wind Power in America's Future*, 13–14.

13. Percentages derived from European Wind Energy Association, *Wind Energy*, 326, table V.1.2.

14. One of the earlier critiques of civilian nuclear power was published under this title. See Richard Curtis and Elizabeth Hogan, *Perils of the Peaceful Atom: The Myth of Safe Nuclear Power Plants* (New York: Doubleday, 1969).

15. See, e.g., Brit Liggett and Jill Fehrenbacher, "Stewart Brand Says Nuclear Power Could Save the World," video interview, Inhabitat.com, February 19, 2011, http://inhabitat.com/.

16. See Adriana Petryna, "The Work of Illness: The Science and Politics of Chernobyl-Exposed Populations," *Osiris*, 2d series, vol. 19, *Landscapes of Exposure: Knowledge and Illness in Modern Environments* (Chicago: University of Chicago Press, 2004), 250–65.

17. See Behrens, *Nuclear Power Plants*, 4.

18. National Academy of Sciences, "Spent Fuel Stored in Pools at Some U.S. Nuclear Power Plants Potentially at Risk From Terrorist Attacks; Prompt Measures Needed to Reduce Vulnerabilities," press release, April 6, 2005. The press release and a link to the full public report are available at http://www8.nationalacademies.org/.

19. Wald, "Resolved," 52.

20. One of those isotopes, plutonium-239, takes 24,000 years to lose half its radioactivity. Plutonium-242 has a half-life of 376,000 years. See *Nuclear Regulatory Commission*, "Backgrounder on Radioactive Waste," April 2007, and "Fact Sheet on Plutonium," October 2003, http://www.nrc.gov/.

21. See U.S. Nuclear Regulatory Commission, "Fact Sheet on Decommissioning Nuclear Power Plants," April 2011, http://www.nrc.gov/.

22. U.S. Energy Information Administration (EIA), *Electric Power Monthly*, table 1.1, accessed July 2011, reporting that petroleum liquids produced 23.4 gigawatt hours of electricity in 2010.

23. In 1973, petroleum products supplied to the U.S. market totaled 17.3 million barrels per day. In 2010, total U.S. oil consumption was 19.1 million barrels per day. U.S. Energy Information Administration, "Short-Term Energy Outlook," March 8, 2011, http://www.eia.doe.gov/.

24. See Pew Center on Global Climate Change, "Transportation Overview: Transportation Emissions in the United States," http://www.pewclimate.org/, accessed March 29, 2011.

25. DONG Energy Annual Report, 2009, 44–45, http://www.dongenergy.com/.

26. Knud Pedersen, vice president responsible for group R&D and regulatory issues, DONG Energy, Copenhagen, interview with author, October 9, 2009. See also Nelson D. Schwartz, "In Denmark, Ambitious Plan for Electric Cars," *New York Times*, December 1, 2009; Clive Thompson, "Batteries Not Included," *New York Times Magazine*, April 16, 2009.

27. EIA, "U.S. Crude Oil, Natural Gas, and Natural Gas Liquids Reserves," table 9, released November 30, 2010, http://www.eia.doe.gov/.

28. In 2010, U.S. consumption of natural gas totaled 24.1 trillion cubic feet. EIA, "Natural Gas Consumption by End Use," released March 29, 2011, http://www.eia.doe.gov/.

29. See, e.g., Natural Gas Supply Association, "Resources," http://www.naturalgas.org/, accessed April 7, 2011. Among the sources cited here is the EIA's *Annual Energy Outlook 2010*, which estimates that the United States has 2,349 trillion cubic feet of "unproved" technically recoverable gas resources.

30. See Sandra Steingraber, "The Whole Fracking Enchilada," *Orion*, September/October 2010, http://www.orionmagazine.org/; Ian Urbina, "Regulation Lax as Gas Wells' Tainted Water Hits Rivers," *New York Times*, February 27, 2011.

31. See Ian Urbina, "Insiders Sound an Alarm Amid a Natural Gas Rush," *New York Times*, June 26, 2011. In the immediate wake of these revelations, congressional leaders called on several federal agencies to investigate the industry's possibly exaggerated claims of shale gas productivity and profitability. See Ian Urbina, "Lawmakers Seek Inquiry of Natural Gas Industry," *New York Times*, June 28, 2011.

32. EIA, "International Gas Reserves and Resources," updated March 3, 2009, http://www.eia.doe.gov/.

33. DOE, *Wind Power in America's Future*, 154, appendix A.

34. Jon Wellinghoff, chairman, Federal Energy Regulatory Commission (FERC), interview with author, Washington, DC, April 15, 2010. Wellinghoff's ideas for developing a smart grid substantially supported by renewable energy are explored further in chapter 9, "Greening the Grid."

35. Pew Center on Global Climate Change, "Renewable & Alternative Portfolio Standards," http://www.pewclimate.org/. This webpage, accessed on March 17, 2011, lists thirty-one states and the District of Columbia with renewable or alternative electricity standards.

36. H.B. 2369, Kan. Reg. Sess. (2009); Kan. State. Ann. Sec. 79–201 (Supp. 2008).

37. Todd Wood, "California Utilities (Just) Miss Renewable Energy Deadline," *Grist*, March 8, 2011, http://www.grist.org/.

38. Yin, "Do State Renewable Portfolio Standards Promote In-State Renewable Generation?" 1144.

39. While not a direct indicator of the cost of a particular energy technology to consumers, wholesale prices—paid by utilities to wind power producers—are generally seen as a more accurate reflection of wind energy's competitiveness than the retail price of power, which can vary widely depending on the generation sources in a utility's portfolio, local dis-

tribution costs, the corporate structure of the utility (public- or investor-owned, regulated or deregulated), and the rate-setting policies of different states.

40. DOE, *2009 Wind Technologies*, 36–42; Ryan H. Wiser, Lawrence Berkeley National Laboratory, "Tracking the U.S. Wind Industry: Update on Cost, Performance, and Pricing Trends," PowerPoint presentation, AWEA Windpower Conference, Anaheim, CA, May 23, 2011, 13–14 (summarizing data forthcoming in the DOE's *2010 Wind Technologies Market Report*).

41. The monthly natural gas price for electric power reached $11.84 per thousand cubic feet in October 2005, dropped as low as $5.76 in 2006, and then spiked again in 2008, topping out at $12.41. In 2010, it dropped to $4.44 in October, but was back up to $5.66 by December. EIA, *Monthly U.S. Natural Gas Electric Power Price*, http://www.eia.gov/.

42. Nancy Rader, California Wind Energy Association, "Wind Energy Development in the Desert," PowerPoint presentation, AWEA Windpower Conference, Anaheim, CA, May 23, 2011.

43. Lazard Ltd., *Levelized Cost of Energy Analysis—Version 4.0, 2010*, cited in National Conference of State Legislatures, *Meeting the Energy Challenges of the Future: A Guide for Policymakers*, 2010, 3, fig. 2, http://www.ncsl.org/.

44. Roland Berger Strategy Consultants, "Innovating China's Wind Energy Market via Engineering Excellence & Supply Chain Integration." PowerPoint presentation, China Wind Power Conference, Beijing, October 21, 2009.

45. Bloomberg New Energy Finance, "Wind Turbine Prices Fall to their Lowest in Recent Years," February 7, 2011, http://bnef.com/.

46. Ibid., 32.

47. DOE, *2009 Wind Technologies Market Report*, 27.

CHAPTER SEVEN: BIRDS AND BATS

1. Center for Biological Diversity v. FPL Group, Inc. et al., Complaint for Violations of California Business and Professions Code Section 17200 et seq., Case No. RG04183113, Superior Court of the State of California, Alameda County, November 1, 2004, paragraphs 46–59.

2. BioResource Consultants, *Developing Methods to Reduce Bird Mortality*, table 1–1.

3. When Audubon reached a legal settlement with Alameda County in 2007, agreeing on a 50 percent mortality reduction goal, the Center for Biological Diversity strongly objected to the agreement's failure to spell out enforceable means of achieving that goal. It registered its views in a letter addressed to the members of the Alameda County Board of Supervisors, January 10, 2007, http://www.biologicaldiversity.org/.

4. Elizabeth McCarthy, "Legal Battle over Altamont Windmill Bird Deaths Ends," *California Current*, December 10, 2010, http://www.cacurrent.com/.

5. Scott Richardson, "Audubon Society Wants Invenergy Study Redone," *Wind Watch: Industrial Wind Energy News*, October 25, 2006, http://www.wind-watch.org/.

6. Dr. Angelo Capparella, associate professor of zoology, Illinois State University, Bloomington, interview with author, May 10, 2009.

7. See National Renewable Energy Laboratory, *Power Technologies Energy Data Book*, Wind Farm Area Calculator, http://www.nrel.gov/, accessed August 1, 2011.

8. Paul Kerlinger, "Prevention and Mitigation of Avian Impacts at Wind Power Facilities" in Resolve, Inc., *Proceedings of the Wind Energy and Birds/Bats Workshop*, 84–85.

9. See Erickson, *Summary and Comparison of Bird Mortality from Anthropogenic Causes*, 1033–34. The Federal Aviation Administration calls for lighting, at a minimum, on towers that define the periphery of all wind farms with turbines taller than 200 feet. See *FAA Advisory Circular: Obstruction Marking and Lighting*, February 12, 2007, chapter 13, http://www.windaction.org/.

10. Capparella interview, May 10, 2009.

11. Karl Kosciuch, "Evaluating Whooping Crane Stop-over Habitat at Potential Wind Power Sites to Understand and Minimize Risk," AWEA Windpower Conference, Chicago, May 7, 2009.

12. Diane Bailey, "Whooping Cranes Use Wind Corridor—Environmental Interests Clash," *Windpower Monthly*, June 1, 2009, 52.

13. Don Furman, senior vice president for development, Iberdrola Renewables, phone interview with author, February 19, 2010.

14. AES Geo Energy, *Saint Nikola Kavarna Wind Farm Environmental Management and Monitoring Plan;* Darius Snieckus, "Early-Warning System Ensures the Safety of Millions of Migratory Birds," *Recharge*, October 8, 2010, http://www.rechargenews.com/.

15. Erickson, *A Summary and Comparison of Bird Mortality from Anthropogenic Causes*, 1035–36.

16. National Research Council, *Environmental Impacts of Wind-Energy Projects*, 75. See also Wallace P. Erickson, "Bird Fatality and Risk at New Generation Wind Projects" in Resolve, Inc., *Proceedings of the Wind Energy and Birds/Bats Workshop*, 29–31.

17. See U.S. Fish and Wildlife Service, *Draft Voluntary Land-Based Wind Energy Guidelines*; and *Draft Eagle Conservation Plan Guidance*. See also *Comments of the American Wind Energy Association on Draft Eagle Conservation Plan Guidance*, May 19, 2011, http://www.awea.org/.

18. Erickson, *Avian Collisions with Wind Turbines*, 7–12.

19. U.S. Fish and Wildlife Service, *Migratory Bird Mortality: Many Human-Caused Threats Afflict Our Bird Populations*, January 2002, http://www.fws.gov/.

20. Nico Dauphine and Robert J. Cooper, "Impacts of Free-Ranging Domestic Cats (*Felis Catus*) on Birds in the United States: A Review of Recent Research with Conservation and Management Recommendations," *Proceedings of the Fourth International Partners in Flight Conference: Tundra to Tropics*, October 2009, 205–9, http://www.abcbirds.org/.

21. The Fish and Wildlife Service avoids giving numerical estimates of past and current sage grouse populations. See U.S. Fish and Wildlife Service, "Questions and Answers for the Greater Sage-Grouse Status Review," 8, website updated March 5, 2010, http://www.fws.gov/. However, when the Interior Department designated the bird as a candidate species under the Endangered Species Act (see discussion later in this chapter), press accounts reported total numbers of sage grouse as ranging from 200,000 to 500,000 today, down from 16 million a century ago. See, e.g., John M. Broder, "No Endangered Status for Plains Bird," *New York Times*, March 5, 2010.

22. Hadassah Reimer, "A Small Bird with a Big Footprint," *Wyoming Lawyer* 31, no. 1 (February 2008): 2, http://www.wyomingbar.org/.

23. State of Wyoming Executive Order 2008–2, Greater Sage-Grouse Area Protection, August 1, 2008, paragraphs 2–3, http://www.fws.gov/. The governor described the purposes of this order in a letter to State Senator Jim Anderson, chair of the legislature's Task Force on Wind Energy, May 18, 2009, http://www.windaction.org/.

24. Freudenthal letter to Anderson, May 18, 2009.

25. For the 2011–12 biennium, mineral severance taxes and federal mineral royalties are projected to account for 51.4 percent of Wyoming's total state revenues. Consensus Revenue Estimating Group, *Wyoming State Government Revenue Forecast, Fiscal Year 2011–Fiscal Year 2016*, October 2010, http://eadiv.state.wy.us/.

26. State of Wyoming Executive Order 2010–4, Greater Sage-Grouse Area Protection, August 18, 2010, Specific Stipulations, http://psc.state.wy.us/.

27. See Wyoming Game and Fish Commission, "Wildlife Protection Recommendations for Wind Energy Development in Wyoming," April 23, 2010, 7–10, http://gf.state.wy.us/.

28. John M. Broder, "No Endangered Status for Plains Bird," *New York Times*, March 5, 2010.

29. Erik Molvar, executive director, Biodiversity Conservation Alliance, Laramie, WY, phone interview with author, January 11, 2010. See also Biodiversity Conservation Alliance, *Wind Power in Wyoming*.

30. Natural Resources Conservation Service and Wildlife Habitat Council, "Greater Prairie-Chicken (Tympanuchus cupido)," *Fish and Wildlife Management Leaflet No. 27*, February 2006, 1–2, http://policy.nrcs.usda.gov/.

31. Robert J. Robel, professor of environmental biology (ret.), Kansas State University, phone interview with author, January 13, 2010.

32. Pitman, "Location and Success of Lesser Prairie-Chicken Nests," 1267.

33. Stephanie Manes, Smoky Hills coordinator, Ranchland Trust of Kansas, phone interview with author, December 23, 2009.

34. Brian Obermeyer, Flint Hills project director, Nature Conservancy, interview with author, July 21, 2010.

35. Rick Plumlee, "Regulations Few in State's Flint Hills Burning Plan," *Wichita Eagle*, December 7, 2010, http://www.kansas.com/.

36. John M. Briggs, director, Konza Prairie Biological Station, Kansas State University, interview with author, July 21, 2010.

37. Robel, "Spring Burning."

38. Rene Braud, director of permitting and environmental affairs, Horizon Wind Energy, phone interview with author, December 17, 2009.

39. See Kunz, "Ecological Impacts of Wind Energy Development on Bats," 316, 320–22.

40. Kunz, "Assessing Impacts of Wind-Energy Development on Nocturnally Active Birds and Bats," 2450. Nationwide, three species of migratory, tree-dwelling bats—the eastern red bat, the hoary bat, and the silver-haired bat—make up almost 75 percent of fatalities from wind turbines. Kunz, "Ecological Impacts of Wind Energy Development on Bats," 316.

41. National Research Council, *Environmental Impacts of Wind Energy Projects*, 97–98.

42. Kunz, "Ecological Impacts of Wind Energy Development on Bats," 318.

43. Arnett, "Patterns of Bat Fatalities," 64.

44. Judge Roger W. Titus, Memorandum Opinion, Animal Welfare Institute, et al., v. Beech Ridge Energy LLC, et al., Case No. RWT 09cv1519, U.S. District Court, District of Maryland, December 8, 2009, 67, http://www.mdd.uscourts.gov/.

45. Judge Roger W. Titus, Order, Animal Welfare Institute, et al., v. Beech Ridge Energy LLC, et al., Case No. RWT 09cv1519, U.S. District Court, District of Maryland, December 8, 2009, http://www.awionline.org/.

46. "Settlement Reached at Beech Ridge Industrial Wind Installation," *Charleston (WV) Daily Mail,* January 27, 2010, http://alleghenytreasures.wordpress.com/.

47. Don Furman, senior vice president for development, transmission, and policy, Iberdrola Renewables, and AWEA president (2009–10), phone interview with author, February 19, 2010. See also "Iberdrola Renewables, BWEC Second Year of Ground-Breaking Bat Study Again Shows Large Reduction in Bat Mortality," press release, November 11, 2010, http://www.iberdrolarenewables.us/.

48. Edward B. Arnett, "Reducing Bat Fatalities at Wind Energy Facilities by Changing Turbine Cut-In Speed," Bat Conservation International, PowerPoint presentation, March 25, 2009, slide 23, U.S. Fish and Wildlife Service website, http://www.fws.gov/.

49. Kunz, "Ecological Impacts of Wind Energy Development on Bats," 316–19.

50. See also U.S. Fish and Wildlife Service, *Draft Voluntary Land-Based Wind Energy Guidelines,* February 2011, http://www.fws.gov/.

CHAPTER EIGHT: THE NEIGHBORS

1. Sylvia White, "Towers Multiply, and Environment is Gone with the Wind," *Los Angeles Times,* November 26, 1984.

2. Paul Gipe, "Aesthetic Guidelines for a Wind Power Future," in Pasqualetti, *Wind Power in View,* 195, photo caption referring to the Zond Victory Garden wind farm in Tehachapi.

3. A revealing account of this controversy appears in Williams, *Cape Wind,* 248–98. See also American Wind Energy Association et al., "Diverse Coalition Announces Campaign to Urge Congress to Reject Backroom Attack on Offshore Wind Power," press release, April 20, 2006, http://www.capewind.org/; Philip Warburg and Susan Reid, "Wind Power with No Direction," *Boston Globe,* February 27, 2006; Philip Warburg and Susan Reid, "Cape Wind Myths and Facts," *Cape Cod Times,* May 16, 2006, http://www.capewind.org/.

4. The rapid return of porpoises to wind farm construction areas was documented at Horns Rev, Denmark's other large offshore wind energy complex. Jakob Tougaard et al., *Harbour Porpoises on Horns Reef—Effects of the Horns Reef Wind Farm,* Annual Status Report 2004 to Elsam Engineering A/S (Roskilde, Denmark: National Environmental Research Institute, July 2005), 7, http://www.hornsrev.dk/.

5. E.ON Sverige, *Rødsand 2 Offshore Wind Farm: Environmental Impact Assessment—Summary of the EIA-Report,* E.On Sverige AB, June 2007, 12, www.dhi.dk.

6. Bjarne Haxgart, site manager, Rødsand 2 Offshore Wind Farm, E.ON Climate & Renewables, Rødbyhavn, Denmark, interview with author, October 5, 2009.

7. E.ON Sverige, *Rødsand 2—Summary of the EIA-Report,* 5, 9.

8. Frank and Sarah Diss, Robert and Ruth Widman, Ransom, IL, interviews with author, May 8, 2009.

9. Invenergy, *Grand Ridge Easement Agreement Summary,* provided to author by Andrew Downey, land agent for the Grand Ridge Wind Farm, May 8, 2009. These fees escalate by either the cost-of-living index or 2 percent annually, whichever is greater.

10. Bob Widman interview, May 8, 2009.

11. Rose Z. Bacon, coproprietor, RK Cattle, Council Grove, KS. This and subsequent quotes are from interviews with author, January 5, 2010, and July 20, 2010.

12. Rose Z. Bacon, "Statement in Support of the Flint Hills National Heritage Area," undated, provided to author by R. Bacon.

13. Ron Klataske, executive director, Audubon of Kansas, phone interview, December 21, 2009. Quotations are from Audubon of Kansas, "Audubon of Kansas is a Leader in Efforts to Protect the Flint Hills," *Prairie Wings*, Fall/Winter 2004, 4–7.

14. Garland P. (Pete) Ferrell III, rancher, Beaumont, KS, and principal, Energy for Generations, LLC, Tulsa. This and subsequent quotes are from interviews with author, January 19 and July 19, 2010.

15. Leslie Wayne, "Brothers at Odds," *New York Times Magazine*, December 7, 1986.

16. Tim Doyle, "Koch's New Fight," *Forbes*, September 21, 2006, http://www.forbes.com/.

17. "Cape Wind Foes Spent $2 Million on Lobbying," *National Journal*, February 23, 2009, http://undertheinfluence.nationaljournal.com/. The multiple steps taken by Koch to fund anti–Cape Wind efforts are documented in Greenpeace, *Bill Koch: The Dirty Money Behind Cape Wind Opposition*, July 22, 2010, http://www.greenpeace.org/.

18. Greenpeace, *Koch Industries: Secretly Funding the Climate Denial Machine* (March 2010) and *Koch Industries: Still Fueling Climate Denial* (2011 rev'd. ed.), http://www.greenpeace.org/. See also Tom Hamburger et al., "Koch Brothers Now at Heart of GOP Power," *Los Angeles Times*, February 6, 2011.

19. See Wind and Prairie Task Force, *Final Report*, 10, fig. 1.

20. See Ken Vandruff, "Governor Considers Making Kansas Flint Hills into a No-Wind-Farm Zone," *Wichita Business Journal*, December 5, 2004, http://www.bizjournals.com/.

21. See Office of the Governor, State of Kansas, "Governor Announces Road Map for Wind Energy Policy," news release, May 6, 2011, https://governor.ks.gov/; Associated Press, "Brownback Strikes Deal with Wind Farms," *Topeka Capital-Journal*, May 8, 2011, http://cjonline.com/.

22. Rose Z. Bacon, e-mail to author, May 6, 2011.

23. Pete Ferrell, Grinnell Alumni Lecture, "Dances with Hooves," June 4, 2010, http://loggia.grinnell.edu/. Ferrell attributes this timeline to Wes Jackson, founder of the Land Institute in Salina, Kansas, and a longtime advocate for sustainable agricultural practices in the prairie.

24. See Gipe, *Wind Power*, 129–46.

25. See Clark, *Energy for Survival*, 541–45; Hills, *Power from Wind*, 274–75; and Righter, *Wind Energy in America*, 126–36.

26. "Mountain-Top Wind Turbine To Be Built by Rutland Corporation," *Rutland (VT) Herald*, January 21, 1946.

27. Nina Keck, "Ira Struggles with Proposed Wind Farm," Vermont Public Radio, June 9, 2009.

28. Vermont Community Wind press releases: "Vermont Community Wind Farm LLC Revises Project Scope," June 18, 2009; "Vermont Community Wind Farm Removes Potential Susie's Peak Turbines from Project Plans," January 15, 2010, http://vtcomwind.com/.

29. Mary Pernal, assistant professor of English, Green Mountain College, Poultney, VT, interview with author, May 6, 2010.

30. Phil Bloomstein, "Living Next to a Wind Turbine," July 1, 2009, personal account, on Industrial Wind Action Group website, http://www.windaction.org/.

31. Wendy Todd, testimony presented to Governor's Task Force on Wind Power Development, State of Maine, September 26, 2007, http://www.maine.gov/. See also Resource Systems Engineering, *Sound Level Study: Ambient & Operations Sound Level Monitoring*, June 21, 2007, http://www.marshillwind.com/.

32. Stanley M. Shapiro, MD, "Can Wind Turbines Cause Heart Disease?" *Heart Health News*, Rutland Regional Medical Center, Winter 2010; "The Industrial Wind Health Impact Debate Rages on at the Rutland (Vermont) Herald," February 6, 2010, on Allegheny Treasures website, http://alleghenytreasures.wordpress.com/.

33. Vermont Public Radio News, "Plan for Ira Wind Is Tabled," April 27 and April 28, 2010, http://www.vpr.net/.

34. Colby, *Wind Turbine Sound and Health Effects.*

35. Brett Horner et al., *Wind Energy Industry Acknowledgment of Adverse Health Effects.* The debate between Drs. McCunney and Nissenbaum, held on May 6, 2010, can be viewed on YouTube, RRMC Wind Health Forum, http://www.youtube.com/.

36. See Colby, *Wind Turbine Sound and Health Effects*, sec. 3.4.1.

37. Among the group living three miles or more away from the wind farm, one person reported new or worsened chronic sleep deprivation, one reported new chronic headaches, and none reported persistent anger or new or worsened depression. See Michael A. Nissenbaum, MD, "Wind Turbines, Health, Ridgelines, and Valleys," media release, May 7, 2010, Society for Wind Vigilance website, http://www.windvigilance.com/. Mars Hill residents have filed a civil lawsuit against the wind farm's owner, First Wind, seeking compensation for emotional and physical distress, as well as for decreased property values. See Jen Lynds, "Mars Hill Windmills Prompt Civil Lawsuit," *Bangor (ME) Daily News*, August 12, 2009, http://new.bangordailynews.com/.

38. See Industrial Wind Action Group website, http://www.windaction.org.

39. Government of Australia, National Health and Medical Research Council, "Wind Turbines and Health," July 2010, http://www.nhmrc.gov.au/.

40. Pierpont, *Wind Turbine Syndrome*, 26, 48–103.

41. Geoff Leventhall, "Direct Testimony on Behalf of Wisconsin Electric Power Company," *Application for a Certificate of Public Convenience and Necessity to Construct and Place in Service a Wind Turbine Electric Generation Facility Known as Glacier Hills Wind Park in Columbia County, Wisconsin*, Docket No. 6630-CE-302, October 20, 2009, R1.80-R1.87, http://www.co.whatcom.wa.us/; and Geoff Leventhall, "Vibroacoustic Disease (VAD) and Wind Turbines," presented as Exhibit 20 at the same proceeding, http://psc.wi.gov/.

42. Nina Pierpont, letter to Geoff Leventhall, January 14, 2007, *Wind Turbine Syndrome* book website, http://www.windturbinesyndrome.com/. In this letter, Pierpont quotes her own remarks at a previous public presentation.

43. See Tech Environmental, "Acoustic Study of Vestas V82 Wind Turbines—Fairhaven, Massachusetts," 4, table 1; and Resource Systems Group, Inc. (RSG), *Noise Impact Study for Georgia Mountain Community Wind*, 4, fig. 2.

44. See RSG, *Noise Impact Study for Georgia Mountain Community Wind*, 6–7.

45. WHO, *Night Noise Guidelines for Europe*, 2009, xiv–xviii.

46. Maine Dept. of Environmental Protection, Chapter 375, "No Adverse Environmen-

tal Effect Standard of the Site Location Law," sec. 10—Control of Noise, effective January 18, 2006, http://www.maine.gov/.

47. Here are a few examples: In Ashe County, North Carolina, wind turbines rated above 20 kilowatts cannot produce more than 5 decibels above the average noise level on adjacent properties and are barred from exceeding 45 decibels except on properties leased or owned by the wind company; Ashe County, North Carolina, *An Ordinance to Regulate Wind Energy Systems,* art. 6, sec. 2.7 (2007). In Chippewa, Minnesota, there is a 50-decibel daytime and nighttime noise limit at adjacent residences, but tighter limits may be imposed on pulsating and tonal sounds; Chippewa, Minnesota, *Windpower Management,* sec. 12.8 (2005). In Huron County, Michigan, turbine noise measured outside residences, schools, hospitals, churches, and public libraries cannot exceed 50 decibels or 5 decibels above ambient levels, whichever is greater, for more than 10 percent of any hour; Huron County, Michigan, *Zoning Ordinance,* art. X, sec. 3E (2009).

48. Colby, *Wind Turbine Sound and Health Effects,* sec. 3.1.3.

49. See Pierpont, *Wind Turbine Syndrome,* 112–21; Colby, *Wind Turbine Sound and Health Effects,* secs. 3–7.

50. Wendy Todd, testimony presented to the Governor's Task Force on Wind Power Development, State of Maine, September 26, 2007, http://www.maine.gov/.

51. National Research Council, *Environmental Impacts of Wind-Energy Projects,* 157–58.

52. Robert Gardiner, cofounder and co-owner, Independence Wind, e-mail to author, February 13, 2011.

53. State of Wisconsin, 2011–2012 Legislature, January 2011 Special Session, Assembly Bill 9, 6–7, http://www.renewwisconsin.org/.

54. RENEW Wisconsin, "Walker Proposal Would Torpedo $1.8 Billion in New Wind Power Investments," January 14, 2011, http://www.renewwisconsin.org/.

55. Public Service Commission of Wisconsin, "Modifications to the Wind Siting Rules Approved by the Public Service Commission on December 9, 2010," PSC Docket 1-AC-231, 3, http://www.renewwisconsin.org/.

56. Jonathan Tilley, "Wisconsin Suspends Wind Siting Rules," *Windpower Monthly,* March 2, 2011, http://www.windpowermonthly.com/.

57. See Affidavit of Michael A. Nissenbaum, MD, State of Maine Board of Environmental Protection, In Re: Record Hill Wind, LLC, September 17, 2009, http://www.wind-watch.org/.

58. See, e.g., State of New Jersey, 214th Legislature, Senate Bill 2374, introduced November 8, 2010, http://www.njleg.state.nj.us/. A stated purpose of this bill, which would establish a 2,000-foot turbine setback from residences and residentially zoned property, is to protect New Jersey residents from "the ill health effects associated with 'wind turbine syndrome,'" whose symptoms it then enumerates. The bill was referred to the Senate Environment and Energy Committee for consideration.

CHAPTER NINE: GREENING THE GRID

1. Bob Whitton, rancher, Bordeaux Junction, WY, and chair, Renewable Energy Alliance of Landowners, interview with author, November 6, 2010.

2. U.S. Department of Energy (DOE), "Wind Powering America: Wyoming Wind Map and Resource Potential," http://www.windpoweringamerica.gov/.

3. The Wyoming Infrastructure Authority has $1 billion in bonding authority for new

transmission infrastructure. See http://wyia.org/. See also Wyoming Statutes 37–5-301 et seq.

4. Victor E. Garber, president, VeJay Energy & Land, Inc., and land agent for Pathfinder Renewable Energy, LLC, e-mail to author, November 19, 2010.

5. See http://www.powercompanyofwyoming.com. See also Associated Press, "Anschutz Corp. Has Wind Farm in the Works," May 25, 2009, http://billingsgazette.com/.

6. Wyoming Infrastructure Authority, Zephyr Project (ZTP), 2011, http://wyia.org/.

7. American Electric Power, "Transmission Facts," undated, 4, http://www.aep.com/, accessed June 19, 2011.

8. Electric Power Research Institute, "EPRI Finds Direct Current Power Uses 15% Less Electricity Than Alternating Current System at Duke Energy Data Center," news release, November 16, 2010, http://my.epri.com/. A higher estimate of 20 percent loss reduction was cited in the summary of the EPRI High Voltage Direct Current & Flexible Alternating Current Transmission Systems Conference, October 28, 2010, http://www.cvent.com/.

9. See Bureau of Land Management, table 1–3: "Mineral and Surface Acres Administered by the Bureau of Land Management, Fiscal Year 2009," http://www.blm.gov/.

10. Pacific Railway Act of 1862 (12 Stat. 489); Pacific Railway Act of 1864 (13 Stat. 356).

11. Walter E. George, national project manager, BLM, Wyoming State Office, Cheyenne, interview with author, November 8, 2010.

12. U.S. Secretary of the Interior, Order No. 3283, Enhancing Renewable Energy Development on the Public Lands, January 16, 2009.

13. Tom Lahti, renewable energy chief, BLM, Wyoming State Office, Cheyenne, interview with author, November 8, 2010.

14. In 2010 alone, the BLM held four public auctions for oil and gas leases in Wyoming, covering more than 300,000 acres. See Bureau of Land Management, *Competitive Lease Sale Notices & Results*, http://www.blm.gov/.

15. Lahti interview, November 8, 2010.

16. See Energy Policy Act of 2005, Public Law 109–58 (August 8, 2005), Sec. 368: Energy Right-of-Way Corridors on Federal Land.

17. See U.S. Departments of Agriculture, Energy, Interior, and Defense, News Release: "Agencies Publish Final Environmental Impact Statement on Energy Corridor Designation in the West," November 26, 2008, http://corridoreis.anl.gov/. These corridors have been legally challenged on a number of grounds by environmental groups. See Kate Galbraith, "Environmentalists Sue over Energy Transmission Across Federal Lands," *New York Times*, July 8, 2009. Western Resource Advocates and fourteen other environmental and conservation groups registered their concerns about the corridors in *Group Comments on the Draft Programmatic Environmental Impact Statement for the Designation of West-Wide Corridors*, February 14, 2008, http://www.westernresourceadvocates.org/.

18. Holtkamp, *Transmission Siting in the Western United States*, 28. This report, in part, describes the state-local division of authority over transmission infrastructure siting in eleven western states.

19. See Wu Qi, "Projects Waiting as Grid Build Lags Behind," *Windpower Monthly, Special Report—China: Market Ambition Ramps Up a Gear*, October 2009, 8–9.

20. See ITC website, http://www.itctransco.com/.

21. Howard Learner, executive director, Environmental Law and Policy Center, phone interview with author, March 18, 2010.

22. These plants include Antelope Valley Station (870 megawatts), Coal Creek Station (1,210 megawatts), Coyote Station (450 megawatts), R.M. Heskett Station (115 megawatts), Leland Olds Station (656 megawatts), and Milton R. Young Station (734 megawatts). See SourceWatch website, http://www.sourcewatch.org/.

23. Federal Energy Regulatory Commission (FERC), Promoting Wholesale Competition Through Open Access Non-Discriminatory Transmission Services by Public Utilities; Recovery of Stranded Costs by Public Utilities and Transmitting Utilities, Order No. 888, Final Rule, April 24, 1996, 75 FERC 61,080, Docket Nos. RM95-8-000 and RM94-7-001, http://www.ferc.gov/.

24. John Rogers et al., *Importing Pollution: Coal's Threat to Climate Policy in the U.S. Northeast,* Union of Concerned Scientists, December 2008, http://www.ucsusa.org/.

25. Edward J. Markey, U.S. rep. (D-MA), Washington, DC, interview with author, April 16, 2010.

26. See Beth Daley, "Markey Urges Federal Cape Wind Approval," *Boston Globe,* November 9, 2009.

27. See, e.g., Wellinghoff, "Federal Transmission Initiatives for Renewables."

28. Jon Wellinghoff, chairman, FERC, interview with author, Washington, DC, April 15, 2010.

29. The Waxman-Markey bill would have required retail electric distributors to provide an increasing percentage of power from renewable sources and electricity-saving measures, reaching 20 percent of all power sold in 2020. See H.R.2454, American Clean Energy and Security Act of 2009, 111th Congress, approved by the House of Representatives, June 26, 2009, Sec. 101, http://www.govtrack.us/. The bill's summary can be found at the same site.

30. See Ballard Spahr Andrews & Ingersoll, LLP, "ARRA Appropriations Provisions Table," http://baltimore.uli.org/.

31. Barack Obama, "Remarks by the President on Clean Energy," Newton, IA, April 22, 2009, http://www.whitehouse.gov/.

32. DOE, National Transmission Grid Study, May 2002, 53–59, http://www.ferc.gov/.

33. See Energy Policy Act of 2005, Public Law 109–58 (August 8, 2005), 16 U.S. Code Annotated Sec. 824p(i)(1), 824p(a), and 824p(b).

34. See Regulations for Filing Applications for Permits to Site Interstate Electric Transmission Facilities, 71 Fed. Reg. 69,440, 69,444 (December 1, 2006).

35. Piedmont Environmental Council et al. v. FERC, 558 F.3d 304 (4th Cir. 2009), cert denied, 130 S.Ct 1138 (2010). An enlightening analysis of this case appears in Dorsi, "Case Comment."

36. American Wind Energy Association, *Green Power Superhighways,* 20–21.

37. Don Furman, senior vice president for development, transmission, and policy, Iberdrola Renewables, and AWEA president (2009–10), phone interview with author, February 19, 2010.

38. Chris Miller, president, Piedmont Environmental Council, Warrenton, VA, interview with author, April 14, 2010.

39. DOE, *Wind Power in America's Future,* 98.

40. See Illinois Commerce Commission et al. v. FERC et al., 576 F.3d 470, 476 (2009).

41. See FERC, *Order Accepting Tariff Revisions,* June 17, 2010, 131 FERC 61,252, Docket No. ER10-1069-000, http://www.ferc.gov/. See also Southwest Power Pool, "FERC Ap-

proves New Cost Sharing Method for Expanding SPP's Transmission Grid," June 17, 2010, http://www.spp.org/.

42. Tierney, *A 21st-Century "Interstate Electric Highway System,"* 44–45.

43. See the discussion of DONG Energy's emerging plug-in vehicle network in chapter 6, "The Path to Cleaner Energy."

44. Wellinghoff interview, April 15, 2010.

45. See Beacon Power, "Smart Energy 25 Flywheel," http://www.beaconpower.com/. See also Business Wire, "Beacon Power Integrates Additional Megawatt of Flywheel Energy Storage on New England Power Grid," December 17, 2009, http://www.businesswire.com/.

46. Throop Wilder, president, 24M Technologies, phone interview with author, February 13, 2011. See also Gregory T. Huang, "A123 Spinoff, 24M Technologies, Raises $10M to Develop Energy Storage Systems for Utilities, Electric Vehicles," Xconomy-Boston, August 16, 2010, http://www.xconomy.com/.

47. Tom Wind, Wind Utility Consulting PC, Jamaica, IA, interview with author, February 4, 2010. See also, "Wrong Geology Sinks Iowa's $400m Wind Storage Dream," *Recharge*, July 29, 2011, http://www.rechargenews.com/.

EPILOGUE

1. See National Priorities Project, "Cost of War," http://costofwar.com/, accessed June 20, 2011.

2. See, e.g., U.S. Partnership for Renewable Energy Finance, "Prospective 2010–2012 Tax Equity Market Observations," July 2010, http://www.uspref.org/.

3. George P. Shultz, "Viewpoints: Clean Air Law Is Key to Our Future," *Sacramento Bee*, September 12, 2010, http://www.sacbee.com/. Proposition 23 sought to freeze implementation of California's Global Warming Solutions Act until the state's unemployment rate dropped to 5.5 percent or below for four consecutive quarters. The state's jobless rate at the time stood at over 12 percent.

Selected Bibliography

Acoustic Ecology Institute. *AEI Special Report: Wind Energy Noise Impacts*, November 17, 2009. http://www.acousticecology.org/.

AES Geo Energy. *Saint Nikola Kavarna Wind Farm—Environmental Management and Monitoring Plan (EMMP)*, November 2008. http://www.aesgeoenergy.com/.

American Wind Energy Association. *An Agenda for the New President and Congress*, November 2008. http://www.newwindagenda.org/.

———. *AWEA Year-End 2009 Market Report*, January 2010. http://www.awea.org/.

———. *AWEA Mid-Year 2010 Market Report*, July 2010. http://www.awea.org/.

———. *U.S. Wind Industry Year-End 2010 Market Report*, January 2011. http://www.awea.org/.

———. *U.S. Wind Industry Annual Market Report—Year Ending 2010*. April 2011. http://www.awea.org/.

American Wind Energy Association, Blue Green Alliance, United Steelworkers. *Winds of Change: A Manufacturing Blueprint for the Wind Industry*, June 2010. http://www.awea.org/.

American Wind Energy Association, Solar Energy Industries Association. *Green Power Superhighways: Building a Path to America's Clean Energy Future*, February 2009. http://www.awea.org/.

America's Energy Future. *Electricity from Renewable Energy Resources: Status, Prospects, and Impediments*. Washington, DC: National Academies Press, 2010.

Anderson, R., J. Tom, N. Neumann, et al. *Avian Monitoring and Risk Assessment at the San Gorgonio Wind Resource Area*. National Renewable Energy Laboratory Subcontract Report NREL/SR-500–38054, August 2005. http://www.nrel.gov/.

Arnett, Edward B., W. Kent Brown, Wallace P. Erickson, et al. "Patterns of Bat Fatalities at Wind Energy Facilities in North America." *Journal of Wildlife Management* 72, no. 1 (2008): 61–78. http://www.batsandwind.org/.

Asmus, Peter. *Reaping the Wind: How Mechanical Wizards, Visionaries, and Profiteers Helped Shape Our Energy Future*. Washington, DC: Island Press, 2001.

Atkinson, Rob, Michael Shellenberger, Ted Nordhaus, et al. *Rising Tigers, Sleeping Giant: Asian Nations Set to Dominate the Clean Energy Race by Out-investing the United States*. Breakthrough Institute and the Information Technology and Innovation Foundation, November 2009. http://www.thebreakthrough.org/.

Awerbuch, Shimon, Leonard S. Hyman, and Andrew Vesey. *Unlocking the Benefits of Restructuring: A Blueprint for Transmission*. Vienna, VA: Public Utilities Reports, Inc., 1999.

Baker, T. Lindsay. *Blades in the Sky: Windmilling through the Eyes of B. H. "Tex" Burdick*. Lubbock: Texas Tech University Press, 1992.

———. *A Field Guide to American Windmills*. Norman: University of Oklahoma Press, 1985.

Barclay, Robert M. R., E. F. Baerwald, and J. C. Gruver. "Variation in Bat and Bird Fatalities at Wind Energy Facilities: Assessing the Effects of Rotor Size and Tower Height." *Canadian Journal of Zoology* 85, no. 3 (March 2007): 381–88.

Behrens, Carl, and Mark Holt. *Nuclear Power Plants: Vulnerability to Terrorist Attack*. Congressional Research Service, Library of Congress, February 2005. http://www.globalsecurity.org/.

Benedetti, Tara. "Running Roughshod? Extending Federal Siting Authority over Interstate Electric Transmission Lines." *Harvard Law Journal on Legislation* 47 (Winter 2010): 253.

Bennett, Aaron. *Integration of Variable Generation Task Force (IVGTF)*. North American Electric Reliability Corporation, March 17, 2009.

Berry, David. "Innovation and the Price of Wind Energy in the U.S." *Energy Policy* 37 (2009): 4493–99.

Biodiversity Conservation Alliance. *Wind Power in Wyoming: Doing It Smart from the Start*, November 2008. http://www.voiceforthewild.org/.

BioResource Consultants. *Developing Methods to Reduce Bird Mortality in the Altamont Pass Wind Resource Area*. California Energy Commission Public Interest Energy Research Program, August 2004. http://www.energy.ca.gov/.

Bird, Lori, Mark Bolinger, Troy Gagliano, et al. "Policies and Market Factors Driving Wind Power Development in the United States." *Energy Policy* 33 (2005): 1397–1407.

Branco, Castelo, and M. Alves-Pereira, "Vibroacoustic Disease." *Noise and Health* 6 (2004): 3–20. http://www.noiseandhealth.org/.

Breukers, Sylvia, and Maarten Wolsink. "Wind Power Implementation in Changing Institutional Landscapes: An International Comparison." *Energy Policy* 35 (2007): 2737–50.

Brown, Ashley C., and Jim Rossi. "Siting Transmission Lines in a Changed Milieu: Evolving Notions of the 'Public Interest' in Balancing State and Regional Considerations." *University of Colorado Law Review* 81 (Summer 2010): 705.

Brown, Brit T., and Benjamin A. Escobar. "Wind Power: Generating Electricity and Lawsuits." *Energy Law Journal* 28, no. 2 (2007): 489–515. http://bmpllp.com/.

Brown, Lester R. *Plan B 4.0: Mobilizing to Save Civilization*. New York: W. W. Norton, 2009.

Buen, Jorund. "Danish and Norwegian Wind Industry: The Relationship Between Policy Instruments, Innovation and Diffusion." *Energy Policy* 34 (2006): 3887–97.

Butler, Lucy, and Karsten Neuhoff. *Comparison of Feed in Tariff, Quota and Auction Mechanisms to Support Wind Power Development*. Cambridge, UK: University of Cambridge Working Papers in Economics, 2004.

Charles River Associates. *SPP WITF Wind Integration Study*. Prepared for Southwest Power Pool, January 4, 2010. http://www.uwig.org/.

Chief Medical Officer of Health. *The Potential Health Impact of Wind Turbines*. Ontario: Queen's Printer for Ontario, May 2010. http://www.health.gov.on.ca/.

Chiras, Dan. *Power from the Wind: Achieving Energy Independence*. Gabriola Island, Canada: New Society Publishers, 2009.

Clark, Wilson. *Energy for Survival: The Alternative to Extinction*. Garden City, NY: Anchor Books, 1974.

Clarke, Alexi. "Wind Energy: Progress and Potential." *Energy Policy* 19, no. 8 (1991): 742–55.

Colby, W. David, Robert Dobie, Geoff Leventhall, et al. *Wind Turbine Sound and Health Effects: An Expert Panel Review.* Prepared for the American Wind Energy Association and the Canadian Wind Energy Association, December 2009. http://www.canwea.ca/.

Conrad, Rebecca, and Susan Hess. *Tallgrass Prairie National Preserve Legislative History, 1920–1996.* Prepared for National Park Service Midwest Support Office, 1998.

Crachilov, Constantin, Randall S. Hancock, and Gary Sharkey. *China Greentech Report 2009.* Mango Strategy, 2009. http://www.china-greentech.com/.

Danish Commission on Climate Change Policy. *Green Energy—The Road to a Danish Energy System Without Fossil Fuels,* September 28, 2010. http://www.klimakommissionen .dk/.

DeBlieu, Jan. *Wind: How the Flow of Air Has Shaped Life, Myth, and the Land.* New York: Houghton Mifflin, 1998.

Dietz, Brian. "Turbines vs. Tallgrass: Law, Policy, and a New Solution to Conflict over Wind Farms in the Kansas Flint Hills." *Kansas Law Review* 54 (2006): 1131–63.

Dorsi, Michael S. "Case Comment: *Piedmont Environmental Council v. FERC.*" *Harvard Environmental Law Review* 34 (2010): 593–603. http://www.law.harvard.edu/.

EnerNex Corporation. *Eastern Wind Integration and Transmission Study.* Prepared for the National Renewable Energy Laboratory, January 2010. http://www.nrel.gov/.

Environmental Law Institute. *Estimating U.S. Government Subsidies to Energy Sources: 2002–2008.* Washington, DC: September 2009.

———. *State Enabling Legislation for Commercial-Scale Wind Power Siting and the Local Government Role.* Washington, DC: May 2011.

Erickson, Wallace P., Gregory D. Johnson, and David P. Young Jr. *A Summary and Comparison of Bird Mortality from Anthropogenic Causes with an Emphasis on Collisions.* USDA Forest Service Technical Report PSW-GTR-191, 2005. http://www.fs.fed.us/.

Erickson, Wallace P., Gregory D. Johnson, David P. Young Jr., et al. *Synthesis and Comparison of Baseline Avian and Bat Use, Raptor Nesting and Mortality Information from Proposed and Existing Wind Developments.* Prepared for Bonneville Power Administration, December 2002. http://www.bpa.gov/.

Erickson, Wallace P., Gregory D. Johnson, M. Dale Strickland, et al. *Avian Collisions with Wind Turbines: A Summary of Existing Studies and Comparisons to Other Sources of Avian Collision Mortality in the United States.* National Wind Coordinating Committee Resource Document, August 2001. http://www.west-inc.com/.

Etherington, John. *The Wind Farm Scam: An Ecologist's Evaluation.* London: Stacey International, 2009.

European Wind Energy Association. *Wind Energy—The Facts,* 2009. http://www .wind-energy-the-facts.org.

Fagan, Mark L. *Understanding the Patchwork Quilt of Electricity Restructuring in the United States.* Regulatory Policy Program Working Paper RPP-2006–04. Cambridge, MA: Mossavar-Rahmani Center for Business and Government, John F. Kennedy School of Government, Harvard University, 2006.

Ferrey, Steven. "Nothing but Net: Renewable Energy and the Environment, *Mid-American* Legal Fictions, and Supremacy Doctrine." *Duke Environmental Law & Policy Forum* 14, no. 1 (2003): 1.

———. "Power Future." *Duke Environmental Law & Policy Forum* 15, no. 26 (Spring 2005): 261–73.

———. "Restructuring a Green Grid: Legal Challenges to Accommodate New Renewable Energy Infrastructure." *Environmental Law* 39 (2009): 977–1014.

Fershee, Joshua P. "269 Levels of Green: State and Regional Efforts, in Wyoming and Beyond, to Reduce Greenhouse Gas Emissions." *Wyoming Law Review* 7 (2007): 269.

Fielder, J. K., T. H. Henry, R. D. Tankersley, and C. P. Nicholson. *Results of Bat and Bird Mortality Monitoring at the Expanded Buffalo Mountain Windfarm, 2005*. Tennessee Valley Authority, June 2007. http://www.tva.gov/.

Fox-Penner, Peter. *Electric Utility Restructuring: A Guide to the Competitive Era*. Vienna, VA: Public Utility Reports, 1998.

Friedman, Thomas L. *Hot, Flat, and Crowded: Why We Need a Green Revolution—And How It Can Renew America*. New York: Farrar, Straus and Giroux, 2008.

GE Energy. *Western Wind and Solar Integration Study*, May 2010. http://www.nrel.gov/.

Gipe, Paul. *Wind Energy Comes of Age*. New York: John Wiley & Sons, 1995.

———. "Wind Energy Comes of Age: California and Denmark." *Energy Policy* 19, no. 8 (1991): 756–67.

———. *Wind Power: Renewable Energy for Home, Farm, and Business*. White River Junction, VT: Chelsea Green, 2004.

Global Insight. *U.S. Metro Economies: Current and Potential Green Jobs in the U.S. Economy*. Prepared for the United States Conference of Mayors and the Mayors Climate Protection Center, October 2008. http://www.usmayors.org/.

Global Wind Energy Council. *Global Wind Energy Outlook 2010*, October 2010. http://www.gwec.net/.

———. *Global Wind 2008 Report*, February 2009. http://www.gwec.net/.

———. *Global Wind 2009 Report*, March 2010. http://www.gwec.net/.

Godby, Rob, and Roger Coupal. "The Impact of Wind Development on Local Economies—Preliminary Wage Findings." Testimony Before the Wyoming Wind Energy Task Force, August 26–27, 2009. http://legisweb.state.wy.us/.

Hadley, Stanton W., and Alexandra A. Tsvetkova. "Potential Impacts of Plug-in Hybrid Electric Vehicles on Regional Power Generation." *Electricity Journal* 22, no. 10 (December 2009): 56–68.

Hagen, Christian A., Brent E. Jamison, Kenneth M. Giesen, et al. "Guidelines for Managing Lesser Prairie-Chicken Populations and Their Habitats." *Wildlife Society Bulletin* 32 (2004): 69–82.

Hagen, Christian A., Brett K. Sandercock, James C. Pitman, et al. "Spatial Variation in Lesser Prairie-Chicken Demography: A Sensitivity Analysis of Population Dynamics and Management Alternatives." *Journal of Wildlife Management* 73, no. 8 (2009): 1325–32.

Han, Jingyi, Arthur P. J. Mol, Yonglong Lu, et al. "Onshore Wind Power Development in China: Challenges behind a Successful Story." *Energy Policy* 37 (2009): 2941–51.

Harris, Coy F., ed. *Windmill Tales*. Lubbock: Texas Tech University Press, 2004.

Hayes, Denis. *Rays of Hope: The Transition to a Post-Petroleum World*. New York: W. W. Norton, 1977.

Heaps, Richard W. *Regional Economic Impact Analysis for the Georgia Mountain Community Wind Project*. Northern Economic Consulting, March 2009. http://psb.vermont.gov/.

Hendricks, Bracken. *Wired for Progress: Building a National Clean-Energy Smart Grid.* Washington, DC: Center for American Progress, February 2009. http://www.americanprogress.org/.

Hills, Richard L. *Power from Wind: A History of Windmill Technology.* New York: Cambridge University Press, 1994.

Hledik, Ryan. "How Green Is the Smart Grid?" *Electricity Journal* 22, no. 3 (April 2009): 29–41.

Hoen, Ben, Ryan Wiser, Peter Cappers, et al. *The Impact of Wind Power Projects on Residential Property Values in the United States: A Multi-Site Hedonic Analysis.* Prepared for the Office of Energy Efficiency and Renewable Energy (Wind & Hydropower Technologies Program), U.S. Department of Energy, December 2009.

Holtkamp, James A., and Mark A. Davidson. *Transmission Siting in the Western United States: Overview and Recommendations Prepared as Information to the Western Interstate Energy Board.* Holland & Hart, LLP, August 2009. http://www.hollandhart.com/.

Horn, Jason W., Edward B. Arnett, and Thomas H. Kunz. "Behavioral Responses of Bats to Operating Wind Turbines." *Journal of Wildlife Management* 72 (2008): 123–32.

Horner, Brett, Richard R. James, Roy D. Jeffery, et al. *Wind Energy Industry Acknowledgement of Adverse Health Effects: An Analysis of the American/Canadian Wind Energy Association Sponsored "Wind Turbine Sound and Health Effects, An Expert Panel Review, December 2009."* Society for Wind Vigilance, January 2010. http://www.wind-watch.org/.

Howe, Robert W., William Evans, and Amy T. Wolf. *Effects of Wind Turbines on Birds and Bats in Northeastern Wisconsin.* Report submitted to Wisconsin Public Service Corporation and Madison Gas and Electric Company, November 2002. http://www.batsandwind.org/.

Hoy, Jim. *Flint Hills Cowboys: Tales of the Tallgrass Prairie.* Lawrence: University Press of Kansas, 2006.

Hyman, Leonard S., Andrew S. Hyman, and Robert C. Hyman. *America's Electric Utilities: Past, Present, and Future.* 8th ed. Vienna, VA: Public Utilities Reports, 2005.

Information Office of the State Council of the People's Republic of China. *China's Energy Conditions and Policies,* December 2007. http://en.ndrc.gov.cn/.

ISO New England. *New England 2030 Power System Study.* Report to the New England Governors, February 2010. http://www.iso-ne.com/.

Jain, Aaftab, Paul Kerlinger, Richard Curry, et al. *Annual Report for the Maple Ridge Wind Power Project, Postconstruction Bird and Bat Fatality Study—2006.* Draft report prepared for PPM Energy and Horizon Energy and the Technical Advisory Committee for the Maple Ridge Project, February 2007. http://www.wind-watch.org/.

James, Christopher A. "Testimony on Behalf of the Sierra Club before the State Corporation Commission of Virginia." PATH Allegheny Virginia Transmission Corporation: Application for Approval of Electric Facilities Under the Utility Facilities Act, Docket No. PUE-2009–00043, October 14, 2009. http://ceds.org/.

Johnson, Anna, and Staffan Jacobsson. *The Emergence of a Growth Industry—A Comparative Analysis of the German, Dutch, and Swedish Wind Turbine Industries.* Draft paper to be presented at the Joseph A. Schumpeter Society conference, University of Manchester, UK, June–July 2000. http://www.cric.ac.uk/.

Johnson, Gregory D., Matthew K. Perlik, Wallace P. Erickson, et al. "Bat Activity, Composition, and Collision Mortality at a Large Wind Plant in Minnesota." *Wildlife Society Bulletin* 32, no. 4 (2004): 1278–88. http://www.windrush-energy.com/.

Johnson, Gregory D., Wallace P. Erickson, M. Dale Strickland, et al. "Mortality of Bats at a Large-Scale Wind Power Development at Buffalo Ridge, Minnesota." *American Midland Naturalist* 150 (2003): 332–42.

Joskow, Paul L. "Lessons Learned from Electricity Market Liberalization." Special issue, *Energy Journal* 29, no. 2 (2008): 9–42.

Kahn, Edward. "The Production Tax Credit for Wind Turbine Powerplants Is an Ineffective Incentive." *Energy Policy* 24, no. 5 (1996): 427–35.

Kammen, Daniel M., Kamal Kapadia, and Matthias Fripp. *Putting Renewables to Work: How Many Jobs Can the Clean Energy Industry Generate?* Report of the Renewable and Appropriate Energy Laboratory, University of California, Berkeley, April 2004. http://rael.berkeley.edu/.

Kansas Energy Council. *Wind Energy Siting Handbook: Guideline Options for Kansas Cities and Counties*, April 2005. http://kec.kansas.gov/.

Kempton, Willett, Jeremy Firestone, Jonathan Lilley, et al. "The Offshore Wind Power Debate: Views from Cape Cod." *Coastal Management* 33 (2005): 119–49.

Kerlinger, Paul. *An Assessment of the Impacts of Green Mountain Power Corporation's Wind Power Facility on Breeding and Migrating Birds in Searsburg, Vermont, July 1996–July 1998*. National Renewable Energy Laboratory, Subcontract Report NREL/SR-500–28591, March 2002. http://www.nrel.gov/.

Kerns, Jessica, and Paul Kerlinger. *A Study of Bird and Bat Collision Fatalities at the Mountaineer Wind Energy Center, Tucker Count, West Virginia: Annual Report for 2003*. Prepared for FPL Energy and Mountaineer Wind Energy Center Technical Review Committee, February 2004. http://www.wvhighlands.org/.

Kirby, Brendan J. *Frequency Regulation Basics and Trends*, Oak Ridge National Laboratory, December 2004. http://www.ornl.gov/.

Klassen, Ger, Asami Miketa, Katarina Larsen, et al. "The Impact of R&D on Innovation for Wind Energy in Denmark, Germany, and the United Kingdom." *Ecological Economics* 54 (2005): 227–40.

Klein, Arne, Anne Held, Mario Ragwitz, et al. *Evaluation of Different Feed-In Tariff Design Options—Best Practice Paper for the International Feed-In Cooperation*. Karlsruhe, Germany: Fraunhofer Institute for Systems and Innovation Research, October 2008. http://www.worldfuturecouncil.org/.

Komanoff, Charles. "Whither Wind: A Journey Through the Heated Debate over Wind Power." *Orion*, September/October 2008. http://www.orionmagazine.org/.

Krapels, Edward N. *Integrating 200,000 MWs of Renewable Energy into the US Power Grid: A Practical Proposal*. Wakefield, MA: Anbaric Transmission, February 2009. http://www.maine.gov/.

Kunz, Thomas H., Edward B. Arnett, Brian M. Cooper, et al. "Assessing Impact of Wind-Energy Development on Nocturnally Active Birds and Bats: A Guidance Document." *Journal of Wildlife Management* 71, no. 8 (2007): 2449–86. http://www.batsandwind.org/.

Kunz, Thomas H., Edward B. Arnett, Wallace P. Erickson, et al. "Ecological Impacts of Wind Energy Development on Bats: Questions, Research Needs, and Hypotheses." *Frontiers in Ecology and the Environment* 5, no. 6 (2007): 315–24. http://www.migrate.ou.edu/.

Kuvlesky, William P., Jr., Leonard A. Brennan, Michael L. Morrison, et al. "Wind Energy Development and Wildlife Conservation: Challenges and Opportunities." *Journal of Wildlife Management* 71, no. 8 (2007): 2487–98.

Landworks. *Aesthetic Assessment of the Proposed Georgia Mountain Community Wind Project.* Prepared for Georgia Mountain Community Wind, LLC, March 2009. http://psb.vermont.gov/.

Larrabee, Aimée, and John Altman. *Last Stand of the Tallgrass Prairie.* New York: Metro Books, 2001.

Lazzari, Salvatore, and Jane Gravelle. *Effective Tax Rates on Solar/Wind and Synthetic Fuels as Compared to Conventional Energy Resources.* Congressional Research Service, Library of Congress, May 1984.

Leddy, Krecia L., Kenneth F. Higgins, and David E. Naugle. "Effects of Wind Turbines on Upland Nesting Birds in Conservation Reserve Program Grasslands." *Wilson Bulletin of Ornithology* 111, no. 1 (March 1999): 100–104. http://elibrary.unm.edu/.

Lema, Adrian, and Kristian Ruby. "Between Fragmented Authoritarianism and Policy Coordination: Creating a Chinese Market for Wind Energy." *Energy Policy* 35 (2007): 3879–90.

Leventhall, Geoff. Vibroacoustic Disease (VAD) and Wind Turbines Critique. Submission to the Wisconsin Public Service Commission on Behalf of Wisconsin Electric Power Company, October 20, 2009. Docket No. 6630-CE-302, PSC Ref. No. 121879, Exhibit 20.

Lewis, Joanna I. *A Comparison of Wind Power Industry Development Strategies in Spain, India, and China.* Prepared for the Center for Resource Solutions, San Francisco, July 19, 2007. http://www.newenergyindia.org/.

———. *A Review of the Potential International Trade Implications of Key Wind Power Industry Policies in China.* Prepared for the Energy Foundation China Sustainable Energy Program, October 2007. http://www.resource-solutions.org/.

Lewis, Joanna I., and Ryan H. Wiser. "Fostering a Renewable Energy Technology Industry: An International Comparison of Wind Industry Policy Support Mechanisms." *Energy Policy* 35 (2007): 1844–57. http://eetd.lbl.gov/.

Li Junfeng, Gao Hu, Shi Pengfei, et al. *China Wind Power Report 2007.* Beijing: China Environmental Science Press, 2007. http://www.greenpeace.org/.

Li Junfeng, Shi Jingli, Xie Hongwen, et al. *A Study on the Pricing Policy of Wind Power in China.* Greenpeace, Global Wind Energy Council, Chinese Renewable Energy Industries Association, October 26, 2006. http://www.greenpeace.org/.

Linden, Eugene. *The Winds of Change: Climate, Weather, and the Destruction of Civilizations.* New York: Simon & Schuster, 2006.

Lovins, Amory B. *Soft Energy Paths: Toward a Durable Peace.* Cambridge, MA: Ballinger, 1977.

Mahan, Simon, Isaac Pearlman, and Jacqueline Savitz. *Untapped Wealth: Offshore Wind Can Deliver Cleaner, More Affordable Energy and More Jobs Than Offshore Oil.* Oceana, September 2010. http://na.oceana.org/.

Mainzer, Elliot. Statement Before the Committee on Energy and Natural Resources, United States Senate, Hearing on the Role of Grid-Scale Energy Storage in Meeting Our Energy and Climate Goals. December 10, 2009. Bonneville Power Administration website: http://www.bpa.gov/.

Manon Kamp, Linda. "Danish and Dutch Wind Energy Policy 1970–2000: Lessons for the Future." *International Journal of Environment and Sustainable Development* 5, no. 2 (2006): 213–20.

Manwell, J. F., J. G. McGowan, and A. L. Rogers. *Wind Energy Explained: Theory, Design, and Application.* Chichester, UK: John Wiley & Sons , 2002.

Massey, Garth. "Critical Dimensions in Urban Life: Energy Extraction and Community Collapse in Wyoming." *Urban Life* 9, no. 2 (July 1980): 187–99. http://jce.sagepub.com/.

McElroy, Michael B., Xi Lu, Chris P. Nielsen, et al., "Potential for Wind-Generated Electricity in China," *Science* 325, no. 5946 (September 2009): 1378–80.

Menz, Frederic C., and Stephan Vachon. "The Effectiveness of Different Policy Regimes for Promoting Wind Power: Experiences from the States." *Energy Policy* 34 (2006): 1786–96.

Meyer, Niels I. "European Schemes for Promoting Renewables in Liberalised Markets." *Energy Policy* 31 (2003): 665–76.

Meyer, Niels I., and Anne Louise Koefoed. "Danish Energy Reform: Policy Implications for Renewables." *Energy Policy* 31 (2003): 597–607.

Miller, Christopher G. Testimony Before the Subcommittee on Energy and the Environment of the Committee on Energy and Commerce, U.S. House of Representatives. Hearing on the Future of the Grid: Proposals for Reforming National Transmission Policy, June 12, 2009.

Mills, Andrew, Ryan Wiser, and Kevin Porter. *The Cost of Transmission for Wind Energy: A Review of Transmission Planning Studies*. Berkeley, CA: Lawrence Berkeley National Laboratory, February 2009.

Minan, John H., and William H. Lawrence. "Encouraging Solar Energy Development Through Federal and California Tax Incentives." *Hastings Law Journal* 32 (1980–81): 1–58.

Munksgaard, Jesper, and Anders Larsen. "Socio-economic Assessment of Wind Power—Lessons from Denmark." *Energy Policy* 26, no. 2 (1998): 85–93.

Munksgaard, Jesper, and Poul Erik Morthorst. "Wind Power in the Danish Liberalised Power Market—Policy Measures, Price Impact, and Investor Incentives." *Energy Policy* 36 (2008): 3940–47.

Musgrove, Peter. *Wind Power*. New York: Cambridge University Press, 2010.

National Research Council. *Environmental Impacts of Wind-Energy Projects*. Washington, DC: National Academies Press, 2007.

National Wind Coordinating Committee. *Permitting of Wind Energy Facilities: A Handbook*, August 2002. http://www.nationalwind.org/.

———. *Wind Power Facility Siting Case Studies: Community Response*, June 2005. http://www.windaction.org/.

Newman, James, and Edward Zillioux. *Comparison of Reported Effects and Risks to Vertebrate Wildlife from Six Electricity Generation Types in the New York/New England Region*. New York State Energy Research and Development Authority, March 2009. http://www.nyserda.org/.

Nissen, Povl-Otto, Therese Quistgaard, Jyette Thorndahl, et al. *Wind Power—The Danish Way: From Poul la Cour to Modern Wind Turbines*. Askov, Denmark: Poul la Cour Foundation, September 2009.

Noor, John. "Herding Cats: What to Do When States Get in the Way of National Energy Policy." *North Carolina Journal of Law and Technology* 11 (Fall 2009): 145.

Office of the Director of Defense Research and Engineering. *Report to the Congressional Defense Committees: The Effect of Windmill Farms on Military Readiness 2006*. U.S. Department of Defense, 2006. http://www.defense.gov/.

Pasqualetti, Martin J., Paul Gipe, and Robert W. Righter, eds. *Wind Power in View: Energy Landscapes in a Crowded World*. London: Academic Press, 2002.

Pattanariyankool, Sompop, and Lester B. Lave. "Optimizing Transmission from Distant Wind Farms." *Energy Policy* 38 (2010): 2806–15.

Pedersen, Eja, Frits van den Berg, Roel Bakker, et al. "Response to Noise from Modern Wind Farms in the Netherlands." *Journal of the Acoustical Society of America* 126, no. 2 (August 2009): 634–43. http://umcg.wewi.eldoc.ub.rug.nl/.

Pedersen, Eja, and Kerstin Persson Waye. "Wind Turbine Noise, Annoyance, and Self-Reported Health and Well-Being in Different Living Environments." *Occupational Environmental Medicine* 64 (March 2007): 480–86. http://ncbi.nlm.nih.gov/.

Pedersen, Eja, and Pernilla Larsman. "The Impact of Visual Factors on Noise Annoyance Among People Living in the Vicinity of Wind Turbines." *Journal of Environmental Psychology* 28 (2008): 379–89. http://www.mfe.govt.nz/.

Percival, S. M. *Assessment of the Effects of Offshore Wind Farms on Birds*. Prepared for DTI Consulting. ETSU W/13/00565/REP, DTI/Pub URN 01/1434, 2001. http://files.zite3.com/.

Petersen, Ib Krag. *Bird Numbers and Distributions in the Horns Rev Offshore Wind Farm Area—Annual Status Report 2004*. Denmark: National Environmental Research Institute, 2005. http://www.hornsrev.dk/.

Pew Charitable Trusts. *Who's Winning the Clean Energy Race? Growth, Competition, and Opportunity in the World's Largest Economies*. Philadelphia: G-20 Clean Energy Factbook, 2010. http://www.pewtrusts.org/.

Philipson, Lorrin, and H. Lee Willis. *Understanding Electric Utilities and De-Regulation*. 2d ed. London: Taylor & Francis, 2006.

Pierpont, Nina. *Wind Turbine Syndrome: A Report on a Natural Experiment*. Santa Fe: K-Selected Books, 2009.

Pitman, James C., Christian A. Hagen, Robert J. Robel, et al. "Location and Success of Lesser Prairie-Chicken Nests in Relation to Vegetation and Human Disturbance." *Journal of Wildlife Management* 69, no. 3 (2005): 1259–69.

Pruett, Christin L., Michael A. Pattern, and Donald H. Wolfe. "It's Not Easy Being Green: Wind Energy and a Declining Grassland Bird." *BioScience* 59, no. 3 (2009): 257–62. http://www.bioone.org/.

Pruitt, Bettye H. *Timken: From Missouri to Mars—A Century of Leadership in Manufacturing*. Boston: Harvard Business School Press, 1998.

Publicover, David. *A Methodology for Assessing Conflicts Between Windpower Development and Other Land Uses*. Appalachian Mountain Club, AMC Technical Report 04–2, May 2004. http://www.outdoors.org/.

Redlinger, Robert Y., Per Dannemand Andersen, and Paul Erik Morthorst. *Wind Energy in the 21st Century: Economics, Policy, Technology, and the Changing Electricity Industry*. New York: Palgrave, 2002.

Reece, Erik. *Lost Mountain: A Year in the Vanishing Wilderness*. New York: Riverhead Books, 2006.

Resolve, Inc. Proceedings of the Wind Energy and Birds/Bats Workshop: Understanding and Resolving Bird and Bat Impacts, Washington, DC, May 18–19, 2004. September 2004. http://www.ewashtenaw.org/.

Resource Systems Group, Inc. *Noise Impact Study for Deerfield Wind, LLC, Scarsburg/Readsboro, Vermont*. Prepared for PPM Energy, December 2006. http://www.sheffieldwind.com/.

———. *Noise Impact Study for George Mountain Community Wind, George and Milton, Vermont.* February 2009. http://www.georgiamountainwind.com/.

———. *Noise Primer.* March 2008. http://www.state.vt.us/.

Righter, Robert W. *Wind Energy in America: A History.* Norman: University of Oklahoma Press, 1996.

Robel, Robert J., John A. Harrington Jr., Christian A. Hagen, et al. "Effect of Energy Development and Human Activity on the Use of Sand Sagebrush Habitat by Lesser Prairie Chickens in Southwestern Kansas." In *Transactions of the 69th North American Wildlife and Natural Resources Conference,* Washington, DC: Wildlife Management Institute, 2004.

Robel, Robert J., John P. Hughes, Scott D. Hull, et al. "Spring Burning: Resulting Avian Abundance and Nesting in Kansas CRP." *Journal of Range Management* 51, no. 2 (March 1998): 132–38.

Roberts, Mark. "Evaluation of the Scientific Literature on the Health Effects Associated with Wind Turbines and Low Frequency Sound." Wood Dale, IL: Exponent, Inc., October 20, 2009. Prepared for Wisconsin Public Service Commission, Docket No. 6630-CE-302, PSC Ref. No. 121885, Exhibit 27. http://www.maine.gov/.

Rosen, Daniel H., and Trevor Houser. *China Energy: A Guide for the Perplexed.* Joint Project of the Center for Strategic and International Studies and the Peterson Institute for International Economics, May 2007. http://www.iie.com/.

Rossi, Jim. "The Political Economy of Energy and Its Implications for Climate Change Legislation." *Tulane Law Review* 84 (2009): 379.

———. "The Trojan Horse of Electric Power Transmission Line Siting Authority." *Environmental Law* 39 (2009): 1015–48.

Schwartz, Marc, Donna Heimiller, Steve Haymes, et al. *Assessment of Offshore Wind Energy Resources for the United States.* National Renewable Energy Laboratory, Technical Report NREL/TP-500–45889, June 2010. http://www.nrel.gov/.

Sijm, J. P. M. *The Performance of Feed-in Tariffs to Promote Renewable Energy Electricity in European Countries.* Energy Center of the Netherlands, November 2002. http://www.ecn.nl/.

Singh, Virinder, Jeffrey Fehrs, and BBC Research and Consulting. *The Work That Goes into Renewable Energy.* Renewable Energy Policy Project, Research Report, November 2001. http://www.repp.org/.

Sissine, Fred. *Renewable Energy: A New National Commitment?* Congressional Research Service, Library of Congress, October 1993. http://fpc.state.gov/.

———. *Renewable Energy: Tax Credit, Budget, and Electricity Production Issues.* Congressional Research Service, Library of Congress, January 2006.

Sissine, Fred, and Michele Passarelli. *Renewable Energy Technology: A Review of Legislation, Research, and Trade.* Congressional Research Service, Library of Congress, March 1987.

Smith, D. R. "The Wind Farms of the Altamont Pass Area." *Annual Review of Energy* 12 (1987): 145–83.

Solar Energy Research Institute. *A New Prosperity: Building a Sustainable Energy Future.* Andover, MA: Brick House Publishing, 1981.

Spath, Pamela L., Margaret K. Mann, and Dawn R. Kerr, *Life Cycle Assessment of Coal-Fired Power Production.* National Renewable Energy Laboratory. NREL/TP-570–25119, June 1999. http://www.nrel.gov/.

Srivastava, Anurag K., Bharath Annabathina, and Sukumar Kamalasadan. "The Challenges and Policy Options for Integrating Plug-In Hybrid Electric Vehicles into the Electric Grid." *Electricity Journal* 23, no. 3 (April 2010): 83–91.

Stahl, Brent, Lisa Chavarria, and Jeff D. Nydegger. "Wind Energy Laws and Incentives: A Survey of Selected State Rules." *Washburn Law Journal* 49 (Fall 2009): 99–142. http://www.washburnlaw.edu/.

Starrs, Thomas A. "Legislative Incentives and Energy Technologies: Government's Role in the Development of the California Wind Energy Industry." *Ecology Law Quarterly* 15 (1988): 103–58.

Sterzinger, George. *Component Manufacturing: Kansas' Future in the Renewable Energy Industry.* Renewable Energy Policy Project, Technical Report, April 2008. http://www.repp.org/.

Svedarsky, W. Daniel, Ronald L. Westemeier, Robert J. Robel, et al. "Status and Management of the Greater Prairie-Chicken *Tympanuchus cupido pinnatus* in North America." *Wildlife Biology* 6, no. 4 (2000): 277–84. http://www.wildlifebiology.com/.

Swanstrom, Debbie, and Meredith M. Jolivert. "DOE Transmission Corridor Designations & FERC Backstop Siting Authority: Has the Energy Policy Act of 2005 Succeeded in Stimulating the Development of New Transmission Facilities?" *Energy Law Journal* 30 (2009): 415–66. http://www.felj.org/.

Swofford, Jeffrey, and Michael Slattery. "Public Attitudes of Wind Energy in Texas: Local Communities in Close Proximity to Wind Farms and Their Effect on Decision Making. *Energy Policy* 38 (2010): 2508–19.

Szarka, Joseph. "Wind Power, Policy Learning, and Paradigm Change." *Energy Policy* 34 (2006): 3041–48.

Tech Environmental, Inc. *Acoustic Study of Vestas V82 Wind Turbines, Fairhaven, Massachusetts.* Prepared for the Town of Fairhaven, May 2007. http://masstech.org/.

Tierney, Susan F. *A 21st-Century "Interstate Electric Highway System"—Connecting Consumers and Domestic Clean Power Supplies.* Prepared for AEP Transmission, October 31, 2008. http://www.puc.nh.gov/.

Toke, Dave. "Wind Power in UK and Denmark: 'Can Rational Choice Help Explain Different Outcomes?'" *Environmental Politics* 11, no. 4 (Winter 2002): 83–100.

Tomain, Joseph P. "'Steel in the Ground': Greening the Grid with the iUtility." *Environmental Law* 39 (2009): 931–76. http://www.elawreview.org/.

U.S. Department of Energy. *"Grid 2030": A National Vision for Electricity's Second 100 Years,* July 2003. http://www.climatevision.gov/.

———. *2008 Wind Technologies Market Report,* July 2009. http://eetd.lbl.gov/.

———. *2009 Wind Technologies Market Report,* August 2010. http://www1.eere.energy.gov/.

———. *Wind Power in America's Future: 20% Wind Energy by 2030.* Mineola, NY: Dover Publications, 2010.

U.S. Energy Information Administration. *Emissions of Greenhouse Gases in the United States–2009,* March 2011. http://www.eia.gov/.

U.S. Fish and Wildlife Service. Briefing Paper: "Prairie Grouse Leks and Wind Turbines: U.S. Fish and Wildlife Service Justification for a 5-Mile Buffer from Leks; Additional Grassland Songbird Recommendations," July 2004. http://www.fws.gov/.

———. *Draft Eagle Conservation Plan Guidance,* dated January 2011; released February 8, 2011. http://www.fws.gov/.

———. *Draft Voluntary Land-Based Wind Energy Guidelines*, released February 8, 2011. http://www.fws.gov/.

———. *Interim Guidelines to Avoid and Minimize Wildlife Impacts from Wind Turbines*. Department of the Interior, May 2003. http://www.fws.gov/.

U.S. General Accounting Office. *Export Promotion: Federal Efforts to Increase Exports of Renewable Energy Technologies*, December 1992. http://www.legistorm.com/.

U.S. House of Representatives. *Comprehensive National Energy Policy Act: Hearing Before the Committee on Ways and Means*. 102nd Cong., 2d sess., April 28, 1992.

———. *Electricity Competition: Hearings Before the Subcommittee on Energy and Power of the Committee on Commerce*. 105th Cong., 1st sess., October 21 and 22, 1997.

———. *H.R. 1216—Renewable Energy and Energy Efficiency Technology Competitiveness Act of 1989: Hearing Before the Subcommittee on Energy Research and Development of the Committee on Science, Space, and Technology*. 101st Cong., 1st sess., May 23, 1989.

———. *Independent Power Producers: Hearing Before the Subcommittee on Energy Conservation and Power of the Committee on Energy and Commerce*. 99th Cong., 2d sess., June 11, 1986.

———. *PURPA: Renewable Energy Programs: Hearing Before the Subcommittee on Energy and Power of the Committee on Energy and Commerce*. 101st Cong., 2d sess., June 14, 1990.

———. *Renewable Energy Development: Hearings Before the Subcommittee on Energy Research and Development of the Committee on Science, Space, and Technology*. 100th Cong., 1st sess., July 8 and 9, 1987.

———. *Renewable Energy Incentives: Hearing Before the Subcommittee on Energy Conservation and Power of the Committee on Energy and Commerce*. 99th Cong., 1st sess., June 20, 1985.

———. *Renewable Energy Industries: Hearing Before the Subcommittee on Energy Conservation and Power of the Committee on Energy and Commerce*. 99th Cong., 2d sess., December 16, 1986.

———. *Renewable Energy in the Eighties: Needs for Further R&D: Hearings Before the Subcommittee on Energy Development and Applications of the Committee on Science and Technology*. 97th Cong., 2d sess., May 28 and July 28, 1982.

———. *Renewable Energy Technologies: Hearing Before the Subcommittee on Energy and Power of the Committee on Energy and Commerce*. 100th Cong., 2d sess., April 27, 1988.

———. *Unlocking America's Energy Resources: Next Generation: Hearing Before the Subcommittee on Energy and Air Quality of the Committee on Energy and Commerce*. 109th Cong., 2d sess., May 18, 2006.

———. *Wind Energy: Hearing Before the Subcommittee on Energy of the Committee on Science and Astronautics*. 93rd Cong., 2d sess., May 21, 1974.

———. *Wind Energy Systems Act of 1980: Hearings Before the Subcommittee on Energy Development and Applications of the Committee on Science and Technology*. 96th Cong., 1st sess., September 18, 24, 26, and October 17, 1979.

U.S. Offshore Wind Collaborative. *U.S. Offshore Wind Energy: A Path Forward*. October 2009. http://www.usowc.org/.

U.S. Senate. *Clean Energy: From the Margins to the Mainstream: Hearing Before the Committee on Finance*. 110th Cong., 1st sess., March 29, 2007.

———. *Energy Efficiency and Renewable Energy Research, Development, and Demonstration: Hearing Before the Subcommittee on Energy Research and Development of the Committee on Energy and Natural Resources*.101st Cong., 1st sess., June 15, 1989.

———. *Energy Tax Incentives: Hearing Before the Subcommittee on Energy and Agricultural Taxation of the Committee on Finance.* 102nd Cong., 1st sess., June 13 and 14, 1991.

———. *Implementation of the Public Utility Regulatory Policies Act of 1978: Hearings Before the Committee on Energy and Natural Resources.* 99th Cong., 2d sess., June 3 and 5, 1986.

———. *National Energy Policy Act of 1989 (Energy Efficiency and Renewable Energy): Hearing before the Committee on Energy and Natural Resources.* 101st Cong., 1st sess., March 14, 1989.

———. *Renewable Electricity: Hearing Before the Committee on Energy and Natural Resources.* 110th Cong., 2d sess., June 17, 2008.

———. *Solar Development Initiative Act of 1987 and the Renewable Energy and Energy Conservation Competitiveness Act of 1987: Hearing Before the Subcommittee on Energy Research and Development of the Committee on Energy and Natural Resources.* 100th Cong., 1st sess., August 6, 1987.

Van Est, Rinie. *Winds of Change: A Comparative Study of the Politics of Wind Energy Innovation in California and Denmark.* Utrecht, Netherlands: International Books, 1999.

Vann, Adam. *Wind Energy: Offshore Permitting.* Congressional Research Service, Library of Congress, September 2009. http://assets.opencrs.com/.

Vattenfall. *Life-Cycle Assessment: Vattenfall's Electricity in Sweden,* January 2005. http://www.vattenfall.com/.

Vestas. *An Environmentally Friendly Investment: Lifecycle Assessment of a V82–1.65 MW Onshore Wind Turbine,* undated. http://www.vestas.com/.

———. *An Environmentally Friendly Investment: Lifecycle Assessment of a V90–3.0 MW Offshore Wind Turbine,* undated. http://www.vestas.com/.

———. *An Environmentally Friendly Investment: Lifecycle Assessment of a V90–3.0 MW Onshore Wind Turbine,* undated. http://www.vestas.com/.

———. *Life Cycle Assessment of Offshore and Onshore Sited Wind Power Plants Based on Vestas V90–3.0 MW Turbines,* June 2006. http://www.vestas.com/.

Wald, Matthew, Garry Brown, Mike Morris, et al. "Resolved: Using Nuclear and Coal Power in an Environmentally Friendly Manner is the Path Forward in Controlling Climate Change." *Environmental Forum* 27, no. 1 (January/February 2010): 46–53.

Wellinghoff, Jon. "Federal Transmission Initiatives for Renewables: The View from FERC." Keynote address at the Renewable Energy World Conference, Austin, TX, February 23, 2010. http://www.ferc.gov/.

———. "Testimony Before the Energy and Environment Subcommittee of the Committee on Energy and Commerce, U.S. House of Representatives." Hearing on the Future of the Grid: Proposals for Reforming National Transmission Policy, June 12, 2009. http://www.ferc.gov/.

Western Electricity Coordinating Council. *Overview of Policies and Procedures for Regional Planning, Project Review, Project Rating Review, and Progress Reports,* April 2005. http://www.wecc.biz/.

Western Renewable Energy Zones. *Western Renewable Energy Zones—Phase 1 Report.* June 2009. http://www.westgov.org/.

White, Sarah, and Jason Walsh. *Greener Pathways: Jobs and Workforce Development in the Clean Energy Economy.* Center on Wisconsin Strategy, University of Wisconsin, Madison, Workforce Alliance, Apollo Alliance, 2008. http://apolloalliance.org/.

Wild Earth Guardians. *Lesser Prairie-Chicken: A Decade in Purgatory,* June 2008. http://www.wildearthguardians.org/.

Wilderness Society. "Attn: Docket Nos. 2007-OE-01 and 2007-OE-02, Re: Proposed Designation of National Interest Electric Transmission Corridors." Comments submitted to the Office of Electricity Delivery and Energy Reliability, July 6, 2007. http://wilderness.org/.

Williams, Wendy, and Robert Whitcomb. *Cape Wind: Money, Celebrity, Class, Politics, and the Battle for Our Energy Future on Nantucket Sound.* New York: Public Affairs, 2007.

Willrich, Mason. *Electricity Transmission Policy for America: Enabling a Smart Grid, End-to-End.* MIT Industrial Performance Center Energy Innovation Working Paper 09–003, July 2009. http://web.mit.edu/ipc/.

Wind and Prairie Task Force. *Wind and Prairie Task Force Final Report.* Kansas Geological Survey Open-file Report 2004–29, June 7, 2004. http://kec.kansas.gov/.

Wind Energy Task Force. *Final Report and Recommendations.* Wyoming Legislative Service Office, November 2009. http://legisweb.state.wy.us/.

Windpower Monthly. "China: Market Ambition Ramps up a Gear." Special issue. October 2009.

———. "Europe 2020: Achieving 12–15% Electricity from Wind Power." Special issue. March 2009.

———. "Growing Strong: China's Wind Energy Expands at Home and Abroad." Special issue. October 2010.

———. "Investing in the U.S.: A Regional Guide to America's Wind Energy Market." Special issue. September 2010.

Wind Turbine Guidelines Advisory Committee. *Preamble and Policy Recommendations.* U.S. Fish and Wildlife Service, October 2009. http://www.fws.gov/.

———. *White Paper.* U.S. Fish and Wildlife Service, October 2008. http://www.fws.gov/.

Wizelius, Tore. *Developing Wind Power Projects: Theory and Practice.* London: Earthscan, 2007.

World Health Organization Europe. *Night Noise Guidelines for Europe,* 2009. http://www.euro.who.int/.

Wyoming Game and Fish Department. *Wildlife Protection Recommendations for Wind Energy Development in Wyoming,* November 17, 2010. http://gf.state.wy.us/.

Wyoming Wind Collector and Transmission Task Force. *Report to the Legislative Task Force on Wind Energy, Transmission Sub-Committee,* October 9, 2009.

Yin, Haitao, and Nicholas Powers. "Do State Renewable Portfolio Standards Promote In-State Renewable Generation?" *Energy Policy* 38 (2010): 1140–49.

Zoellner, Tom. *Uranium: War, Energy, and the Rock That Shaped the World.* New York: Viking, 2009.

Index

Acciona, 40, 188
AES Corporation, and AES Geo Energy, 120
Agassi, Shai, 106
Agriculture Department. *See* U.S. Department of Agriculture
air pollution: in China, 61, 63; from coal, 51–52, 63, 99–101; from prairie burning in Kansas, 127–28, 142, 168; from wind compared to other power sources, 100–101. *See also* carbon emissions; greenhouse gas emissions
Alliance to Protect Nantucket Sound. *See* Cape Wind
Altamont Pass, California, 30, 116–17, 118, 119, 134
American Clean Energy and Security Act (2009), 171
American Gas Association, 109
American Recovery and Reinvestment Act (2009), 50
American Superconductor, 39
American Wind Energy Association (AWEA), 51, 80, 93, 152, 159, 172–73
Andersen, Lars, 61
anemometer, 88
Animal Welfare Institute, 130
Anschutz, Philip F., and Anschutz Corporation, 164–65
Antelope Valley Station, North Dakota, 168–69, 171
Arab oil embargo, vii, 22, 25, 26, 109, 180
Audubon: Bay Area, 116; Illinois, 117–18; Kansas, 8, 142, 193
availability (wind turbines), 87
avian impacts of wind farms, bats: and causes of death, 129–30; Indiana bats, x, 130–31; and mitigation efforts, 131–32; and mortality rates, 129, 132
avian impacts of wind farms, birds: bald and golden eagles, 116, 117, 121, 133; and comparative causes of death, 121–22; and Endangered Species Act, 122, 130–31; long-tailed ducks, 137; migratory birds, 119; prairie chicken habitats, x, 9, 11, 125–29, 142; raptor deaths and mitigation efforts, 116–19; sage grouse habitats, x, 122–25, 159; and turbine technology changes, 118–19; white storks, 120; whooping cranes, 119–20
AWEA. *See* American Wind Energy Association (AWEA)

Bacon, Rose and Kent, 140–42, 147–49
balancing wind with other power sources, 105–6, 176–78, 181, 182
Barnhart Crane & Rigging, 80–83
bats. *See* avian impacts of wind farms, bats
Beacon Power Corporation, 177, 179
bearings: magnetic, for flywheel power storage, 177; maintenance of, in wind turbines, 45, 46; Timken and, 52–57, 59, 69, 184
Beaver Ridge Wind Project, Maine, 151
Beecher, Brad, 161
Beech Ridge Energy LLC, 130–31
Beech Ridge Wind Farm, West Virginia, 130–31
Better Place, 106–7
Biodiversity Conservation Alliance, 124–25, 159
biomass energy, 111
birds. *See* avian impacts of wind farms, birds
blades: composition of, 21–22, 31, 149; control-

ling pitch of, 86, 150; defects in and repair of, 31, 47–48, 91, 150; detection of, by bats, 129; length of, 14, 49, 74, 114; length of, related to power output, 114; lower and upper reach of, 118–19; manufacture of, 19–22, 40, 46, 60; shadow flicker created by, 155; sounds produced by, 153, 156–57; swept area of, 114; tip speed of, 153; transportation of, 74–78
BLM. *See* U.S. Bureau of Land Management (BLM)
Bloomstein, Phil, 151, 156
Blue Canyon Wind Farm, Oklahoma, 8
Bode, Denise, 93
Boeing, 31
Bonus company, 32
Boots, Mick, 41
Bousbib, Ari, 48
BP Wind Energy, 165
Bradford, Peter, 103
Brand, Stewart, 101
Braud, Rene, 128–29
Briggs, John, 128
Brightwell, Joe, 89
Brown, Jerry, 28–30, 34, 101, 183
Brownback, Sam, 148
Browning, Bill, 143
Buffalo Mountain Wind Energy Center, Tennessee, 129
Bush, George W., viii, 124, 183
Butler, Barry, 89

California: bird deaths from wind turbines in, 116–18, 121; Danish turbine exports to, 24–25, 32–35, 114; electric vehicle network planned in, 107; energy policy in, 28–35, 183–84; hydroelectricity in, 111; need for out-of-state renewable power in, 165; public utility wind power purchases in, 28, 30, 34; renewable electricity standards for, 111, 112, 183–84; solar energy in, 29, 30, 111, 165; tax credits for wind industry in, 30–31, 33–35; visual chaos of early wind farms in, 31, 34, 116, 134–35; wind resources and wind energy use in, vii, 2, 25–35, 183–84
California Energy Commission, 30, 116
California Office of Appropriate Technology, 29
capacity factor (wind power), 3, 97, 114. *See also* installed capacity
cap-and-trade, viii, 56, 169, 171
Cape Wind: Alliance to Protect Nantucket Sound and, 145–46; East Coast perspective on, 18; Ted Kennedy's opposition to, viii, 135–36, 145, 170; Bill Koch's opposition to, 145–46; Ed Markey and, 170; Ted Stevens and, 135–36; Don Young and, 135–36
Capparella, Angelo, 117–19, 124
carbon dioxide (CO_2) emissions. *See* carbon emissions; greenhouse gas emissions
carbon emissions: cap-and-trade regimes for, viii, 56, 169, 171; CO_2 gap between coal, gas, and wind power, 98–101; proposed federal tax on, 171, 183; reduction of, competing with other values, 173; Timken opposition to U.S. regulation of, 52, 56–57. *See also* greenhouse gas emissions
Cardinal Fastener, 39, 80, 184
Cargill, Alvin, 14–16, 79, 80
Carter, Jimmy, vii, viii, 26–28, 34, 109
Casselman Wind Power Project, Pennsylvania, 131–32
Center for Biological Diversity, 116
Charmley, James E., 54–55
Chernobyl nuclear disaster, 63, 101–2, 104, 181
China: air pollution in, 61, 63; coal production and use in, 63, 65; economy of, 62, 64; electricity use in, 62, 68; energy planning and policy in, 62–66, 72; environmental movement in, 67; export of wind turbines from, 60, 68–72; foreign investment in wind farms in, 66–68; hydroelectricity in, 63–64; land acquisition for wind farms in, 65–68; manufacture of wind turbines in, 21, 55–56, 59–62, 68–72; market for U.S. wind energy equipment in, 55–56, 59; nuclear power in, 63; protectionist rules of, 59, 60, 68–69; state and private ownership of wind industry in, 66, 67–68; transmission needs in, 168, 170; Vestas factories in, 60–62, 69, 71; wind resources and wind energy use in, ix, 19, 59, 61–62, 64–65, 187; World Trade Organization and, 69

China Wind Power conference, 59, 60
Chinese Renewable Energy Industries Association, 60
Chokecherry–Sierra Madre wind complex, 164–65
Chunhua Li, 70
Clark, Wilson, 29
climate change: China's contribution to, 63; consequences of, 132–33; fossil fuels and, viii; skepticism about, 90, 142, 146, 173; wind energy as response to, xi, 13, 18, 57, 184. *See also* carbon emissions; greenhouse gas emissions
Clipper Wind: Cedar Rapids, Iowa, and, 39–43; employees and workplace morale at, 40–45; financial problems of, 39, 47–50, 57; lessons to be learned from, 49–52, 57–58; Liberty turbine manufactured by, 42, 44; market share of, 188; quality control issues at, 47–48; remote monitoring by, 44–46; supply chain for, 46–47; United Technologies Corporation bailout of, 48, 49; worker safety at, 45–46
CloudCorp, 10, 17–18
Cloud County, Kansas. *See* Kansas
Cloud County Community College, 17, 87–88, 91–92
coal: air pollution from, 51–52, 63, 99–101; capping carbon emissions from, 56; in China, 63, 65; cost of coal plants, 113; in Denmark, 22–23; energy balance for, 100; failure to internalize environmental costs of, 51, 110; landscape impacts of, 52, 63, 123; mining hazards from, 51–52, 63; price uncertainty of, 112; U.S. policies favoring, 27, 33, 34, 51; U.S. reliance on, 56–57, 96; Wyoming politics regarding sage grouse and, 123–24
coal-bed methane, 108
Colorado: wind manufacturing in, 37, 40; wind resources and wind energy use in, 185
Commonwealth Edison, 70
compressed-air energy storage. *See* storage of electricity
conservation easements, 126–27, 143
Conservation Law Foundation, viii, 135
construction of wind farms, 13–16, 77, 78–84, 90–91, 94, 113
cost: of coal, gas, nuclear, and solar energy vs. wind, 105, 112–13; of modernizing the national grid, 174–76; of wars, 180. *See also* economics of wind power
Crude Oil Windfall Profits Tax (1980), 28
Crum, Lorrie, 53, 56–57
Cube rule of wind power, 114
Culik, Martin, 81
cut-in speed (wind turbines), 131–32

Dalian Heavy Industry Group, 71
Danish International Development Agency, 35
Danish Oil and Natural Gas. *See* DONG Energy
Danish Wind Industry Association, 32
Defense Department. *See* U.S. Department of Defense
Denmark: Arab oil embargo and, 22–23; coal use in, 22–23; DONG (Danish Oil and Natural Gas) in, 105–6, 177; electric vehicles in, 105–7, 177; nuclear plants banned by, 22; offshore wind farms in, 36, 136–38; oil and natural gas in, 22, 23; policies promoting wind in, 23–25, 36; policies reducing greenhouse gas emissions in, 19, 23, 106–7; population of, 37; Risø National Laboratory in, 23, 32; wind energy use in, viii, ix, 19, 22–25, 36–37, 106, 136–38, 187; wind manufacturing in, 32–33, 35–37, 58, 188. *See also* Vestas Wind Systems
Deukmejian, George, 34
DeWind, 188
direct-drive turbines, 55, 70. *See also* gear-driven turbines
Diss, Frank and Sarah, 139
DOE. *See* U.S. Department of Energy (DOE)
Doing It Smart from the Start (Biodiversity Conservation Alliance), 124–25
DONG Energy, 105–6, 177
Donghai Bridge offshore wind farm, China, 65
Dunham, Richard, 89–90

economics of wind power: benefits to rural communities, 3–10, 15–18, 43, 44, 90–91; California tax credits, 30, 33; feed-in tariffs

(Europe), 35–36; capital outlays for wind power, 113; cost of turbines, 11–12, 113; life-cycle costs of wind compared to other power sources, 113; payments to landowners, ix, xi, 3–7, 15–16, 139, 145; productivity gains through technology changes, 113–15; utility purchases of wind power, 11, 12–13, 30, 34, 70, 161–62; wholesale price of wind power, 112. *See also* employment related to wind power; wind power incentives, federal; wind power incentives, state

EDP Renewables, 12

electricity: China's use of, 62, 68; from coal, 23, 56–57, 96; Denmark's shift to wind for, viii, ix, 19, 22–25, 36–37, 106, 136–38, 187; global supply of, by wind, 187–88; from hydroelectric dams, 63–64, 96, 111, 176, 182; from natural gas, 96, 107; number of households supplied with, from U.S. wind farms, ix, xii, 3, 81; present and potential price of, 112–15; purchase of, from U.S. wind farms, 11–13, 27–28, 30, 34, 35–36, 70, 111, 112, 161–62, 165, 183–84; renewable standards for, by states, 111, 112, 165, 183; supplied by wind (U.S.), ix, x, 1, 2, 94, 96, 97–98, 110, 112, 181, 185–86; Timken Company's use of, 56–57

electric vehicles, 106–7, 177, 182

Elk River Wind Power Project, Kansas, 148, 149, 161, 162

Empire District Electric, 12, 161–62

employment related to wind power: in construction, 13–16, 77, 78–84, 90–91, 94; "indirect" and "induced" jobs, 16, 95; in manufacturing, 20, 35, 37, 39–40, 51, 54, 94; in operations and maintenance (O&M), 84–87, 90–91; total jobs (present and projected), 37, 51, 94–95; training programs for technicians, ix, 17, 41, 44, 87–94; women in, 13–15, 17, 79, 82, 91, 92–94

Endangered Species Act. *See* avian impacts of wind farms, birds

Enel, 150

Energias de Portugal (EDP), 12

energy balance, 100

Energy Department. *See* U.S. Department of Energy

Energy Policy Act (1992), 50

Energy Policy Act (2005), 167, 172

Energy Tax Act (1978), 28

Engel, Ditlev, 37

Engelder, Carole, 13–15, 79, 91

environmental issues. *See* air pollution; avian impacts of wind farms, bats; avian impacts of wind farms, birds; carbon emissions; climate change; greenhouse gas emissions; noise from wind farms; visual appearance of wind farms; wildlife impacts of wind farms

Environmental Law and Policy Center, 168, 169, 173

Environmental Protection Agency. *See* U.S. Environmental Protection Agency (EPA)

E.ON Climate and Renewables, 137

EPA. *See* U.S. Environmental Protection Agency (EPA)

European Union: greenhouse gas reduction goal, 106

FAG, 55

farms and farming: in China, 66; economic strains of, ix, 3, 4, 6, 90, 143–44; exodus of young people from, ix, 3–4, 7; in Illinois, 118; in Iowa, 44, 140; in Kansas, ix, 3–5, 9, 12–13, 15–18, 140–49; roots of Vestas turbines in equipment for, 20, 32; water-pumping windmills and, 1–2; wildlife impacts of, 122; wind farms as income supplement to, ix, xi, 3–7, 15–16, 44, 139, 145, 148; and working landscape of farms and ranches compatible with wind, 5, 9, 18, 118, 138–40, 141, 162–64. *See also* landowners

Federal Aviation Administration, 10, 119

Federal Communications Commission, 11

federal energy policy: of George W. Bush, viii, 94–95; of Jimmy Carter, vii, viii, 26–28, 34, 109; on coal, 27, 33, 34, 51; of Richard Nixon, 25–26, 109; of Barack Obama, 39, 50–51, 56, 103, 124, 172, 183; of Ronald Reagan, vii, 34. *See also* coal; natural gas; nuclear energy; oil; solar energy; wind power incentives, federal

Federal Energy Regulatory Commission (FERC), 110, 169–74, 181

feed-in tariffs (Europe), 35–36
FERC. *See* Federal Energy Regulatory Commission (FERC)
Ferrell, Garland P. "Pete," III, 143–49, 157–58, 161
Fish and Wildlife Service. *See* U.S. Fish and Wildlife Service
Fitz, Rusty, 83
Flint Hills, Kansas, x, 8–9, 127, 140–49
flywheel power storage. *See* storage of electricity
Fowler Ridge Wind Farm, Indiana, 45
France: nuclear fuel reprocessing in, 103; wind power use in, 187
Freudenthal, Dave, 123, 124
Friedman, Thomas, 72–73
Fuhrländer, 71
Fukushima Daiichi, Japan, reactor disaster, 52, 63, 102, 104, 180–81
Furman, Don, 131–32, 172–73

Gamesa, 35, 37, 69, 188
Garber, Victor E., 164
Gardiner, Robert, 158
Gates, Robert, 48, 51, 52
gear-driven turbines: assembly of gearboxes in, 41–44, 46, 51; contrast with direct-drive turbines, 55, 70; defects in and repair of, 31, 47–48; monitoring and maintenance of, 45–47, 86–87, 91; quality control in manufacture of, 46–47, 49; rotor-to-generator speed ratio in, 45, 53, 55; supply chain for, 46, 51
General Electric (GE), 31, 39, 40, 45, 46, 59, 69, 71, 86, 151, 188
geothermal energy, 28, 111
Germany: compressed-air energy storage in, 178; feed-in tariff and other policies favoring solar and wind in, 35–36; manufacture of wind turbines in, 35, 58, 188; wind energy use in, viii, 36, 187
GE Wind. *See* General Electric (GE)
Gipe, Paul, 30, 134
Glass, Tyler, 44
global warming. *See* climate change
Global Wind Energy Council, 60
Goldman Sachs, 12, 49
Goldwind Science & Technology. *See* Xinjiang Goldwind Science & Technology
Gordon, Jim, 134
Gore, Al, 142
Graf, Kristen, 93
Graham, Bruce, 17, 87–88, 91
Graham, Michelle, 17, 91
Grandpa's Knob, Vermont, 149–51
Grand Ridge Wind Farm, Illinois, 79, 85–87, 90–91, 138–40
Grassley, Chuck, 50
Gray County Wind Energy Center, Kansas, 3, 146
Great Lakes: manufacturing region for wind, 39; shipment of wind equipment via, 77–78; wind resources of, 57, 98
greenhouse gas emissions: California's agenda on, 184; coal's contribution to Denmark's, 23; Denmark's commitment to reducing, 19, 105–6; European Union commitment to reducing, 106; nuclear power and, 101, 104; Regional Greenhouse Gas Initiative (U.S. Northeast), viii, 169; U.S. as world leader in per capita, 180; U.S. resistance to international curbs on, viii, 19, 146, 180. *See also* carbon emissions; climate change
Greenlight Energy Resources, 146, 147
Greenpeace, 60
Green Power Express, 166, 168–69
green power option. *See* wind power incentives, state
Grumman Aerospace, 31
Guo, Wilson, 66
Guo Zheng, 66

Hansen, H. S., 20
Hansen, Peder, 20
Hartman, Adam, 79
Harvard University, 62
Hawaii, 107, 186
Haxgart, Bjarne, 137
Hayes, Denis, 27
Hemami, Ahmad, 88–89
High Plains Express, 166
Hinkels and McCoy, 79–80, 83
Horizon Wind Energy, 12–16, 79, 81, 85, 113, 126, 161, 163, 165

Hot, Flat, and Crowded (Friedman), 72–73
households supplied by wind farms (U.S.), ix, xii, 3, 81
Hoy, Jim, 8
hubs: attachment of blades to, 49, 82; height of, 49, 114; manufacture of, 42, 46, 61; raising of, 82–83; rotational speed of, 45; transport of, 77
hydropower: in China, 63–64; power storage using, 176, 182; in U.S., 96, 111

Iberdrola Renewables, 120, 131–32, 148
Illinois: Chinese-owned wind farm in, 70; eagle flyway near White Oak Wind Energy Center, 117–18; employment in wind manufacturing in, 40; Grand Ridge Wind Farm in, 85–87, 90–91, 138–40; nuclear energy in, 138; renewable electricity standards in, 111; wind resources and wind energy use in, 57 Illinois State University, 117–18
Independence Wind, 158
India: wind energy use in, 35, 187; wind turbine manufacturing in, 37–38, 188
Indiana: Fowler Ridge Wind Farm in, 45; Meadow Lake Farm in, 74–84, 90; wind resources and wind energy use in, 57, 81
Indiana bats. *See* avian impacts of wind farms, bats
Industrial Wind Action Group, 154
infrasound. *See* noise
installed capacity (wind power): in China, 66; defined, 3; in Denmark, 36, 106, 136–37; in Germany, 36; global, 189; of top ten countries, 187–88; of states, 185–86; of turbines, 13, 31, 99; U.S. total, 36, 181, 187; of U.S. wind farms, xii, 3, 92, 130, 149, 164
Interior Department. *See* U.S. Interior Department
International Atomic Energy Agency, 103
International Organization for Standardization (ISO), 99
Invenergy, 79, 85–86, 117–18, 130, 163
investment tax credits. *See* wind power incentives, federal; wind power incentives, state
Iowa: factory closures in, 40, 41, 42–43; flooding in, 40–41; Iowa Lakes Community College in, 88–90, 92; population of, 39; wind resources and wind energy use in, 185. *See also* Clipper Wind
Iowa Lakes Community College, 88–90, 92
ISO. *See* International Organization for Standardization (ISO)
ITC Transmission, 168

Japan: Fukushima Daiichi nuclear disaster in, 52, 63, 102, 104, 180–81
Jessen, Leo, 85–87, 90–91
Jiquan Wind Farm Base, China, 65
jobs. *See* employment related to wind power
Johnson, Paul, 90

Kansas: Cloud County Community College in, 17, 87–88, 91–92; early uses of wind power in, 1–2; economic benefits of wind power for Cloud County, 16–18; Flint Hills wind controversy in, x, 8–9, 140–49; land deals for wind farms in, 5–16, 144–46; prairie chickens in, 125, 128, 142; ranching practices in, 127–28, 143–44; whooping cranes in, 120; wind energy policy in, 8, 111, 146–48; wind farms in, ix, 3–18, 64, 79, 80, 146–49, 161; wind resources and wind energy use in, 1, 9, 97, 144–45, 149, 185
Kansas Livestock Association, 126, 127
Kansas State University, 93, 126, 128
Kaplan, Matthew, 71–72
Kennedy, Edward M. "Ted," viii, 135–36, 145, 170
Kerlinger, Paul, 119
Klataske, Ron, 142–43
Koch, Bill, 145–46, 183–84
Koch, Charles and David, 146, 183–84
Kocher, Kurt, and family, 3–5, 12–13
Koch Industries, 145–46
Konza Prairie Biological Station, Kansas, 128
Kosciuch, Karl, 120
Kruse, Peter Wenzel, 36–37, 121, 122
Kyoto Protocol, 180

Lahti, Tom, 167
Lalley, Matt, 43, 44
landowners: income from wind farms, ix, xi, 3–7, 15–16, 139, 145, 148; negotiations

with wind developers, 5–16, 144–46, 150, 163–64; perspectives on wind farm noise, 5, 151–59; opposition to wind farms among, 8–9, 140–49, 150–59
land used by wind farms: acreage leased or optioned to wind developers, 6, 10, 12–13, 15, 145, 150, 163–64; footprint of wind farms, 81, 164; "go-zones" in Wyoming, 125; land needed per turbine, 5; setbacks from residential buildings, 157–59; spacing between turbines, 118–19
LaSalle County Nuclear Station, 138
Lawrence Berkeley National Laboratory, 112
layout of wind farms. *See* visual appearance of wind farms
Lazard Ltd., 113
Learner, Howard, 168–69, 173
Leventhall, Geoff, 155
Liberty turbines. *See* Clipper Wind
Li Fan, 70
life cycle environmental assessment, wind and other power sources, 98–101
Li Junfeng, 60
Lone Star Transportation, 74
Lovins, Amory, 29, 30
Lowell, Kirk, 10, 17–18
low-frequency sound. *See* noise
Loyd, Bob, 41–45, 47, 89

Mackinaw Ecosystem Partnership, 117
Maine: Governor's Task Force on Wind Power, 151, 157; wind farm noise in, 151, 154, 156–58; wind resources and wind energy use in, 186
maintenance of wind turbines, 84–87, 90–91
Manes, Stephanie, 126–28
manufacture of wind turbines: in China, 21, 55–56, 59–62, 68–73; in Denmark, ix, 13–14, 19–25, 32–33, 35–37, 58; employment in, 20, 35, 37, 39–40, 94; in Germany, 35, 58; in India, 38; leading manufacturers serving U.S. market, 188; quality control in, 31–33, 46–49, 52–55; smaller companies in, 49–50, 57; in Spain, 35, 37, 58, 69; in U.S., 31–32, 37–58, 188; worker safety and, 45–46; workplace morale and, 44–45. *See also* manufacturers of wind turbine components; manufacturers of wind turbines
manufacturers of wind turbine components. *See* American Superconductor; Cardinal Fastener; FAG; SKF; Timken; TPI Composites; Trinity Structural Towers
manufacturers of wind turbines. *See* Acciona; Bonus; Clipper Wind; DeWind; Fuhrländer; Gamesa; General Electric (GE); Mitsubishi; NEG Micon; Nordtank; REpower; Siemens; Sinovel; Suzlon; United Technologies Corporation (UTC); Vestas Wind Systems; XEMC Windpower; Xinjiang Goldwind Science & Technology
Maples, Steve, 80–84, 90, 91
Markey, Edward J. "Ed," 169–71
Mars Hill wind farm, Maine, 151, 154, 157
Mason, Ray, 5–7, 17
Massachusetts: flywheel power storage in, 177; Ed Markey on wind, 169–71; wind resources and wind energy use in, 97, 186. *See also* Cape Wind
McCunney, Robert, 152–54
McDonnell Douglas, 31
McElwee, Charles R., II, 67
Meader, Mark, 40–41, 43
Meadow Lake Farm, Indiana: construction of, 77, 78–84, 90, 91; delivery of turbine parts to, 74–78; generating capacity of, 81; land used by, 81
Meridian Way Wind Farm, Kansas: conservation easements and, 126–27, 143; construction of, 13–16, 78–80; cost of, 11–12; economic benefits of, for community, 16–18; financial backing for, 12, 49; generating capacity of, ix, 64; income for landowners from, 4–7, 15; jobs created by, 16–17; land acquisition for, 9–16, 127; location of, 3; naming of, 11; prairie chickens at, 126–28; purchase of power from, 11, 12–13, 161–62; transmission lines from, 161–62; Vestas wind turbines at, xii, 3, 4–6, 14, 84–85, 99–100
meteorological towers, 9, 10, 124, 145–46
methane emissions: wind compared to other power sources, 101

Middle East: Arab oil embargo, vii, 22, 25, 26, 109, 180; gas resources in, 109; oil from, 104, 105; political changes in, 180
Miller, Chris, 173–74
Mitsubishi, 38, 188
Molvar, Erik, 124–25
Mortensen, Lone, 19–22

nacelle: described, 14, 42, 44; failed climb attempt, 84; installation of, 80, 81, 82, 83; manufacturing of, 37, 42, 44, 61, 70; transport of, 74; working conditions in, 86, 88–89, 92
NASA, 31
National Academy of Sciences, 102
National Health and Medical Research Council, 154
National Historic Trails, 167
National Renewable Energy Laboratory (NREL), 96–99, 119, 144, 162, 163
National Research Council, 158
natural gas: changing federal policies regarding use of, 33–34; China and, 63; environmental impacts of, 107–10, 122; federal land leases for extraction of, 167; federal subsidies for, 51; FERC jurisdiction over natural gas pipelines, 172; levelized cost of combined-cycle gas plants, 113; life-cycle air emissions of electricity generated by, 99–101; ongoing role for, 181; prospects for shifting to wind power from, 109–10; unconventional recovery methods for (hydraulic fracturing), 108–9; U.S. and foreign supplies of, 107–8; use of, for electricity and household heating, 96, 107; volatile price of, 112; Wyoming sage grouse politics and, 122–25
Nature Conservancy, 8, 143
NEG Micon, 20
Nevada: proposed nuclear waste repository in, 103; transmission line siting in, 165, 167; wind resources and wind energy use in, 97, 186
Newbold, Clint, 80, 82
New York: wind resources and wind energy use in, 57, 81, 185
New Zealand, 35
NextEra Energy Resources, 117
Nissenbaum, Michael, 152, 153–54, 157, 159
nitrogen oxide emissions: wind compared to other power sources, 101
Ni Xiangyu, 61
Nixon, Richard, 25–26, 109, 184
noise from wind farms: annoyance thresholds for, 153–54; distance from turbines and, 151–59; downwind and upwind rotors and, 156; health concerns about, 152–59; setbacks as safeguard against, 157–59; setting noise limits for, 155–60; sound frequencies and, 153
Nordtank company, 32, 35
North China Electric Power University, 66
North Dakota: Antelope Valley Station and coal use in, 168–69; coal piggybacking on Green Power Express in, 168–69; wind resources and wind energy use in, 57, 185
NREL. *See* National Renewable Energy Laboratory (NREL)
nuclear energy: "carbon neutrality" and "clean energy" claims for, 101, 104, 171; Carter's policy on, 27, 33; in China, 63; cost of, 113; decommissioning of, 103; Denmark's ban on, 22; disasters involving, 52, 63, 101–2, 104, 115, 180–81; in France, 103; in Illinois, 138; liability limits and other federal support for, 52, 57; Nixon's policy on, 25; number of nuclear power plants in U.S., 52; poorly matched to wind power balancing needs, 176; Reagan's policy on, vii, 34; share of U.S. electricity production, 96; waste disposal for, 52, 103–4; weapons proliferation risk of, 102–3

Obama, Barack, 39, 50–51, 56, 69, 103, 124, 172, 183
"Obama Bolt," 39, 80
offshore wind power: Cape Wind, viii, 18, 134–36, 145–46, 170, 181; China and, 65; Denmark and, 36, 136–38; Great Lakes and, 57; porpoises and, 137; U.S. potential for, 98; visual appearance of, 134–38
Ohio: wind equipment manufacturing in, 51, 52–57; wind resources and wind energy use in, 57, 186

oil: Arab embargo of, vii, 22, 25, 26, 109, 180; in Denmark, 22, 23; drilling for, on federal land, 167; electric vehicles to reduce reliance on, 104–7; lobbying by leaders of, to block climate change policies, 145–46; Nixon's policy on, 25–26, 109; price history of, 33, 35, 112; Reagan's policy on, 34; subsidies for, 51; U.S. reliance on foreign sources of, 18, 26, 104–7, 180; U.S. consumption of, 26, 96, 104, 105, 107–10; wars fought over, xi, 18, 110–11, 180; Wyoming politics regarding sage grouse protection and, 123–25
Oklahoma: Blue Canyon Wind Farm in, 8; whooping crane migration through, 120; wind resources and wind energy use in, 185
OPEC, 109, 180. *See also* Arab oil embargo
operations and maintenance (O&M) jobs, 84–87, 90–91
opposition to wind farms: in California, 116–17, 134; in Flint Hills, x, 8–9, 140–49; in Vermont, 150–59; in Wyoming, 122–25. *See also* avian impacts of wind farms, bats; avian impacts of wind farms, birds; Cape Wind; noise from wind farms; visual appearance of wind farms
Oregon: wind resources and wind energy use in, 183, 185
Organization of Petroleum Exporting Countries (OPEC), 109, 180. *See also* Arab oil embargo
Oxbow Power Corporation, 145–46

Pacific Gas & Electric, 30, 34
Parsons, Stan, 144
Pathfinder Renewable Energy LLC, 164–65
Pederson, Knud, 105–6, 177
Peñascal Wind Farm, Texas, 120
Pennsylvania: bat protection efforts at Casselman Wind Power Project in, 131–32; wind resources and wind energy use in, 57, 185
Percy, Charles "Chuck," vii
Pernal, Mary, 151
Peyton, Joel, 44, 45–46
Piedmont Environmental Council, Virginia, 172, 173–74
Pierpont, Nina, 154–55, 157, 159
PJM Interconnection, 174–75
Powerplant and Industrial Fuel Use Act (1978), 33
PPM Energy, 147–48
prairie chickens. *See* avian impacts of wind farms, birds
Price Anderson Act, 52
production tax credit (PTC). *See* wind power incentives, federal
Protect the Flint Hills, 148
PTC. *See* production tax credit
Public Utility Regulatory Policies Act (PURPA), 27–28
Putnam, Palmer Cosslett, 149–50

ranches and ranching. *See* farms and farming
Ranchland Trust of Kansas, 126–27, 143
Reagan, Ronald, vii, 34, 184
REAL. *See* Renewable Energy Alliance of Landowners (REAL)
Regional Greenhouse Gas Initiative (RGGI), viii, 169
Reisky, Sandy, 146–48
Ren Dongming, 64–65
renewable electricity standards: adopted by states, 111, 112, 165, 183; proposed federal, 111
Renewable Energy Alliance of Landowners (REAL), 163–64
RENEW Wisconsin, 158
REpower, 188
research and development (R&D): on power storage, 176–78; on wind turbines, 23, 31–35
Revere, Paul, 60
RGGI. *See* Regional Greenhouse Gas Initiative (RGGI)
Risø National Laboratory, Denmark, 23, 32
Robel, Robert, 125–28
Roberts, Jim, 7–16, 66, 127
Rødsand Offshore Wind Farm, Denmark, 136–38
Roggenkamp, Loma, 92–93
Romney, Mitt, 135, 136
Roosevelt, Franklin D., x
rotors: axis of, 31; cut-in speed of, 131–32; expanding diameter of, 114; monitoring speed of, 86; number of blades on, 2, 14,

20, 32, 80, 82–83, 134; rotational speed of, 45, 119, 131–32, swept area of, 114; tip speeds of, 153
Royko, Mike, 29
Rutland Regional Medical Center, 152

safety. *See* worker safety
sage grouse. *See* avian impacts of wind farms, birds
Salazar, Ken, 124
Sammons Enterprises, 164
San Gorgonio Pass, California, 30, 134
Savory, Allan, 144
Sawyer, Steve, 60
Schwarzenegger, Arnold, 183
Sebelius, Kathleen, 8, 146–48
setbacks for wind turbines, from residential buildings, 157–59
"shadow flicker" from wind turbines, 155
Shultz, George, 184
Siemens/Siemens Energy, 38, 40, 92, 188
Sinovel, 71
siting of wind farms: in China, 66–67; in general, 159–60; in Illinois, 117–18, 138–40; Indiana, 75; in Kansas, 5, 8–9, 127, 140–49; in Vermont, 149–59; in Wyoming, 163–64. *See also* avian impacts of wind farms, bats; avian impacts of wind farms, birds; Cape Wind; noise impacts of wind farms; visual appearance of wind farms; wildlife, impacts from wind farms
SKF, 55
smart grid. *See* transmission of electricity
Soft Energy Paths (Lovins), 29
solar energy: in California, 29, 30, 111, 165; Carter's policy on, 26, 27, 34; cost of, 113, 182; definition of, 27; in Germany, 35; Nixon's policy on, 25; solar thermal power, 113; statistics on, 96; tax credit for, 28; White House solar water-heating panels, viii, 27, 34
Solar Energy Industries Association, 172
Solar Energy Research Institute, 27
SourceWatch, 168
Southern California Edison, 30, 34
Southwest Power Pool, 162, 175
Spain: wind energy in, viii, 36, 187; wind industry in, 12, 35, 37, 40, 69, 120, 188
Sporer, Bonnie, 15–16
Stark, Pete, 30
Stevens, Ted, 135–36
St. Nikola Kavarna wind farm, Bulgaria, 120
storage of electricity: batteries and, 178–79, 182; compressed air and, 178; electric vehicles and, 106–7, 177, 182; flywheels and, 177–79, 182; hydro dams and, 176, 182; smart-metered buildings and appliances and, 177
Stovall, Bill, 74–78
sulfur dioxide emissions, 100–101
Sundgren, Jacque and Steve, 143
Supreme Court, U.S. *See* U.S. Supreme Court
Suzlon, 188
swept area. *See* blades; rotors

Tehachapi Pass, California, 30, 31, 134
Tennessee Valley Authority, 129
Texas: drilling for natural gas in, 108; employment in wind manufacturing in, 40; greater prairie chicken in, 125; whooping cranes in, 120; wind energy use in, 183; wind farms in, 120; wind resources and wind energy use in, 185
Three Mile Island nuclear accident, 63, 101, 104, 181
Tianrun New Energy Investment Company, Ltd., 67, 68
Tiedeman, Mary, 44
Tierney, Susan, 175–76
Timken, Henry, 53
Timken, Ward J. "Tim," Jr., 56
Timken Company, 52–57, 59, 69, 184
tip speed. *See* blades
Titus, Roger W., 130–31
Todd, Wendy, 151, 156, 157
Top of the World Windpower Project, Wyoming, 92
towers: buckling of, 31; climbing of, 84–86; erection of, 14, 78–82, 91; height and weight of, ix, 14, 79, 149; height of, related to output, 114; manufacture of, 20, 37, 40, 46, 172; recycling of, 100; transport of, 14, 74, 77–78; visual appearance of, 114, 134; working conditions in, 84, 86, 91–92
TPI Composites, 40

training programs for wind energy technicians, 17, 87–93
TransCanada's Zephyr line, 166
transmission of electricity: to California, 165; in China, 168, 170; citizen groups and, 168–74; coal piggybacking and, 171; costs of, and cost-sharing formulas for new transmission lines, 174–76; East Coast and, 169–70; in Europe, 170; federal role in, 166–74; "smart grid" management tools for, x, 176–78, 182; in Kansas, 161–62; from Midwest, 166, 168–70, 174–75; modernization of, 171–79, 182; in Nevada, 165, 167; and siting of new interstate transmission lines, 161–74; West-Wide Energy Corridor, 167; for wind energy, 161–79; in Wyoming, 164–67. *See also* storage of electricity
transportation: cost of, for turbine transport, 105; electric vehicles for wind power storage and use, 106–7, 177, 182; energy consumed in U.S. for, 105; trucking of turbine components, 14, 74–77; use of rail and ships for turbine transport, 77–78
TransWest Express, 165, 166, 167
Trinity Structural Towers, 40
Tsinghua University, 62
turbines. *See* wind turbines
24M, 178, 179

Union of Concerned Scientists, 169
Union Pacific Railroad, 78, 166
United Kingdom: wind resources and wind energy use in, 35, 187, 188
United Steelworkers, 69
United Technologies Corporation (UTC), 48
University of Georgia, 122
University of Iowa, 89
UPC Renewables, 66, 67, 68
U.S. Bureau of Land Management (BLM), 166–67
USDA. *See* U.S. Department of Agriculture
U.S. Department of Agriculture, 15
U.S. Department of Defense, 10–11
U.S. Department of Energy (DOE), x, 94–95, 100, 109, 110, 121, 160, 172, 175, 181
U.S. Environmental Protection Agency (EPA), 156
U.S. Fish and Wildlife Service, 121, 122, 131, 133
U.S. Interior Department, 122, 124, 166
U.S. Nuclear Regulatory Commission, 103
U.S. Supreme Court, 28, 172
utility purchases of wind power: Commonwealth Edison, in Illinois, 70; Empire District Electric and Westar, in Kansas, 11–13, 161–62; feed-in tariffs as catalysts to, in Europe, 35–36; Nevada and California utilities as prime markets for, 30, 34, 165; PURPA's creation of market for, 27–28; renewable electricity standards as catalysts to, 111, 112, 183–84

Van Beusekom, Justin, 84–85
Van Est, Rinie, 32
variability of wind, and need for balancing, 105–7, 176–78, 181, 182
Vattenfall, 99
Vermont: Grandpa's Knob wind turbine in, 149–51; opposition to wind farms in, 150–59; Taconic Mountains in, x; wind resources and wind energy use in, 186
Vermont Community Wind, 150–52
Vestas Wind Systems: blade manufacture in Lem, Denmark, 19–22; in China, 60–62, 69, 71; and climb attempt, 84; compared with Clipper Wind, 45, 46; and construction of Meadow Lake Wind Farm, Illinois, 74, 77, 80; diagram and dimensions of turbines made by, xii, 14, 74, 84–85; employment by, 20, 35, 37; factory locations of, 20–21, 35, 37, 60–62; global headquarters of, 36–37; history of, 20, 32–36; life-cycle analyses of wind turbines by, 99–100; markets for, 24–25, 35–36, 188; revenues of, 20; as world's top-ranked turbine supplier, ix, 20
Vickerman, Michael, 158
Virginia: as home of Piedmont Environmental Council, 172–74; wind resources and wind energy use in, 186
Visceral Vibratory Vestibular Disturbance, 155
visual appearance of wind farms: aviation

lights on turbines, 137–38; Cape Wind and, 134–36; Flint Hills objections to, 8, 141–43; Illinois farm families' divergent perspectives on, 138–40; increasing size of turbines, 113–15, 119, 135; Kurt Kocher and, 5; layout of turbines and, 15, 81, 116, 118–19, 134–35, 138; National Historic Trails and, 167; Rødsand Offshore Wind Farm (Denmark) and, 136–38; spacing between turbines, 118–19, 138; turbine design changes and, 31, 114, 134–35, 138; visual chaos at older California wind farms, 31, 34, 116, 134–35; water-pumping windmills, 1–2

Walker, Scott, 158–59
Walters, Jeanna, 93–94
water-pumping windmills, 1–2
Waxman, Henry, 171
Wellinghoff, Jon, 110, 170–71, 177, 181
Wennberg, Jeff, 152
Westar Energy, 12, 13, 161
Westinghouse, 31
West Virginia: bat protection effort in, 130–31; wind resources and wind energy use in, 186
West-Wide Energy Corridor, 167
Wheatley, Dave, 41, 43
White, Sylvia, 134
White Oak Wind Energy Center, Illinois, 117–18
white storks. *See* avian impacts of wind farms, birds
Whitton, Bob, 162–65
WHO. *See* World Health Organization (WHO)
Whole Earth Catalog, 101
whooping cranes. *See* avian impacts of wind farms, birds
Widman, Bob and Ruth, 139
Wilder, Throop, 178
wildlife, impacts from wind farms: on big-game species, 125; offshore wind farms, 137; and pacing of wind farm development, 118. *See also* avian impacts of wind farms, bats; avian impacts of wind farms, birds
Wind, Tom, 178
Wind and Prairie Task Force (Kansas), 8
wind chargers, 2
wind farm developers. *See* Beech Ridge Energy LLP; BP Wind Energy; EDP Renewables; Enel; E.ON Climate and Renewables; Greenlight Energy Resources; Horizon Wind Energy; Iberdrola Renewables; Independence Wind; Invenergy; NextEra Energy Resources; Oxbow Power Corporation; Pathfinder Renewable Energy; PPM Energy; Sammons Enterprises; Tianrun New Energy Investment Company, Ltd.; Vermont Community Wind
wind farms (non–U.S.). *See* Donghai Bridge offshore wind farm, China; Jiquan Wind Farm Base, China; Rødsand Offshore Wind Farm, Denmark; St. Nikola Kaverna Wind Farm, Bulgaria
wind farms (U.S.). *See* Altamont Pass, California; Beaver Ridge Wind Project, Maine; Beech Ridge Wind Farm, West Virginia; Blue Canyon Wind Farm, Oklahoma; Buffalo Mountain Wind Energy Center, Tennessee; Cape Wind, Massachusetts; Casselman Wind Power Project, Pennsylvania; Chokecherry–Sierra Madre wind complex, Wyoming; Elk River Wind Power Project, Kansas; Fowler Ridge Wind Farm, Indiana; Grandpa's Knob, Vermont; Grand Ridge Wind Farm, Illinois; Gray County Wind Energy Center, Kansas; Mars Hill wind farm, Maine; Meadow Lake Wind Farm, Indiana; Meridian Way Wind Farm, Kansas; Peñascal Wind Farm, Texas; San Gorgonio Pass, California; Tehachapi Pass, California; Top of the World Windpower Project, Wyoming; White Oak Wind Energy Center, Illinois
wind power capacity: as affected by wind speed ("cube rule"), 114; in California, 30–33, 185; in China, 19, 59–73, 187; as a factor of blade length, 114; global installed wind power capacity (1996–2010), 189; in Kansas, 1–18, 97, 144–45, 149, 185; in Massachusetts, 97, 186; ranking of installed wind power capacity, by country, 187–88;

ranking of U.S. onshore wind power capacity (present and potential), by state, 185–86; in U.S., 96–98, 125, 185–86; U.S. offshore wind power potential, 98, 170, 181; in Wyoming, 125, 162–65, 185
Wind Power in America's Future (DOE report), 94
wind power incentives, federal: grid modernization grants, 172; investment tax credits, 28, 34–35, 51; loan guarantees, 172; Obama stimulus measures, 50–51, 56, 172; production tax credit, 2–3, 13, 50–51, 111, 113, 182–83; technical training grants, 88. *See also* wind power incentives, state
wind power incentives, state: grants and loans, 111; green power option, 111; investment tax credits and tax-free bonds in California, 30–31, 33–35; net-metering laws, 111; property tax exemption in Kansas, 2, 147; renewable electricity standards, 111, 112, 165, 183. *See also* wind power incentives, federal
wind speed: and bat fatalities, 131–32; definition of "windy land areas," 97; impact on turbine capacity factor, 97, 114; impact on turbine durability, 53–54; turbine tower height and, 114; and U.S. wind power potential, 96–98, 185–86
wind turbine components. *See* blades; hubs; nacelles; rotors; towers
wind turbines: availability of, 87; capacity factor of, 3, 97, 114; climbing of, 84–85, 91, 92; cost of, 11–12, 113; cut-in speed for, 131–32; design of, 31–32, 34, 39, 113–17, 149–50; diagram of, xii; dimensions and weight of, xii, 14, 49, 74, 75, 84–85, 92, 114–15, 119; direct-drive vs. gear-driven, 31, 41–51, 53, 55, 70, 86–87, 91; erection of, 13–14, 81–84, 90–91; foundations for, 80; installed capacity of, 13, 31, 99; land required for, 5, 118–19; maintenance of, 84–87, 90–91; noise from and noise limits for, x, 5, 149–60; research and development on, 23, 31–35; rotational speed of, 119; setbacks for, from neighboring homes, 157, 158–59; "shadow flicker" from, 155; transport of, 74–78; wildlife impacts from, x, 116–33, 159, 167. *See also* manufacture of wind turbines; visual appearance of wind farms; wind turbine components
Wind Turbine Syndrome, 154–55
Wisconsin: turbine setback controversy in, 158–59; wind resources and wind energy use in, 185
women, employment of: at wind farms, 13–15, 17, 79, 82, 91–94; in technical training programs, 91–94; in top corporate positions, 93
Women of Wind Energy, 93
Wood, Grant, 18
worker safety: in operations and maintenance of turbines, 83, 86, 88, 92–93; at turbine manufacturing plants, 45–46
World Health Organization (WHO), 156
World Trade Organization (WTO), 69
Wyoming: coal mining in, 122, 123; coalition of pro-wind landowner associations (REAL) in, 163–64; electric transmission lines originating in, 164–67; federal lands in, 166–67; National Historic Trails in, 167; oil and gas resources in, 123–24; ranching in, 162–63; sage grouse in, 122–25, 159; wildlife conservation in, 122–23, 125, 159, 167; wind farm siting in, 123–25, 167; wind resources and wind energy use in, 124–25, 162–65, 185
Wyoming Infrastructure Authority, 164

XEMC Windpower, 55
Xinjiang Goldwind Science & Technology, 67–68, 70–71

Young, Don, 135
Yucca Mountain, Nevada, 103

Zehtindjiev, Pavel, 120
Zephyr transmission line, 164–65, 166, 167
Zilkha Renewable Energy, 7, 8–12, 49, 127, 147